The Economics of Climate Resilience in Power Infrastructure

IEEE Press
445 Hoes Lane
Piscataway, NJ 08854

The Economics of Climate Resilience in Power Infrastructure

Ali Arabnya
Quanta Technology, USA
University of Denver, USA

Amin Khodaei
University of Denver, USA

IEEE Press
WILEY

Published by John Wiley & Sons, Inc., Hoboken, New Jersey.

Published simultaneously in Canada.

For general information on our other products and services or for technical support, please contact our Customer Care Department within the United States at (800) 762-2974, outside the United States at (317) 572-3993 or fax (317) 572-4002.

Wiley also publishes its books in a variety of electronic formats. Some content that appears in print may not be available in electronic formats. For more information about Wiley products, visit our web site at www.wiley.com.

Library of Congress Cataloging-in-Publication Data applied for:

Hardback ISBN: 9781394220755

Cover Design: Wiley
Cover Image: © Anton Petrus/Getty Images

Set in 9.5/12.5pt STIXTwoText by Straive, Chennai, India

To our families

Contents

About the Authors

Ali Arabnya, PhD, is the Director of Infrastructure Finance & Climate Risk with Quanta Technology based in Washington, DC. He is also a research professor of electrical and computer engineering with the University of Denver. Previously, he was a consultant climate economist with the World Bank. Prior to that, he was a Data & Analytics Manager with Protiviti in New York City. He has published over 40 peer-reviewed technical papers and is the co-author of the book *The Economics of Microgrids*. He is a Senior Member of the IEEE Power & Energy Society and holds a PhD in Industrial Engineering from the University of Houston.

Amin Khodaei, PhD, is a Professor of Electrical and Computer Engineering at the University of Denver. His research is focused on the climate crisis, the grid of the future, and grid-enabling technologies including artificial intelligence and quantum computing. He has published over 220 peer-reviewed technical articles on various aspects of electric grid modernization. He is a Senior Member of the IEEE and holds a PhD in Electrical Engineering from the Illinois Institute of Technology.

Foreword

Electrical power and energy systems are critical to power the prosperity of our society. They form the backbone of modern economy and are central to every aspect of our lives. Present-day electrical systems are confronted with unprecedented challenges, such as weather-related impacts and physical and cybersecurity threats. They are also enablers of the decarbonization of the environment through deployment of renewable generation, electrification of transportation and heating, and other sources of carbon emissions.

Power system resilience is critical to achieving society's goals by adapting to and mitigating climate change and is a cornerstone of societal stability and prosperity. Adaptation means designing and operating resilient energy infrastructure that can withstand and quickly recover from the shocks of climate change, while mitigation includes reducing the greenhouse gas emissions that are driving these extreme events.

This book addresses adaptation challenges to ensure that our power systems remain resilient, reliable, and sustainable in the face of climate change and helps with achieving resilience goals. The authors provide a comprehensive overview and analysis of the engineering economics of climate adaptation in electric power infrastructure. Through a series of developed models, case studies, and real-world applications, the authors present strategies to enhance the resilience of our power systems, focusing on the critical role of economic decision-making in managing climate risks. From understanding the economic implications of climate shocks to developing new financial instruments, asset management frameworks, and operational strategies, this book lays out a blueprint for building more resilient and adaptive power systems.

Each chapter provides both theoretical insights and practical solutions through comprehensive framework that can guide engineers, policymakers, and utilities in the critical task of future-proofing our power systems. Whether it is through advanced optimization techniques, innovative insurance models, or existing and new technologies, the solutions presented here emphasize the importance of a holistic, interdisciplinary approach to resilience.

The importance of resilience in power systems cannot be overstated. As we continue to witness the increasing frequency and intensity of climate-related disruptions, the ability of our energy infrastructure to quickly recover—and to adapt to new risks—will determine the well-being of communities, economies, and societies worldwide. This book provides invaluable tools and strategies for meeting this challenge.

In an age where the need for sustainability is an immediate necessity, this book stands as a guide to ensuring that our electrical power systems are resilient, adaptable, affordable, and prepared for the opportunities and challenges ahead.

December 05, 2024

Dr. Damir Novosel
President of Quanta Technology

Preface

Climate change is one of the grand challenges of the 21st century. It poses an increasing risk to our critical infrastructure and requires significant investment for upgrades, mitigation, and adaptation. Power grid infrastructure in particular is expected to be highly stressed by these events, combined with an ever-increasing power demand, reduced operational capacity, and increased probability of system failures resulting in significant power outages. The economic and financial losses caused by climate-related events such as wildfires, heatwaves, and hurricanes can reach to a level that threatens the financial health and solvency of the utilities and local governments. In addition, they can significantly hinder the ability of utilities and infrastructure owners to recover, which can lead to dire consequences for shareholders, ratepayers, and policymakers. That requires developing a wide range of innovative engineering economic solutions to achieve the financial, operational, and physical resilience expected from a 21st-century electric power infrastructure. The economics of resilience should be incorporated in technological solutions to streamline the success of climate adaptation measures in the power and energy industry. However, despite significant engineering research and development efforts on climate adaptation in power infrastructure, there is a lack of concerted effort to provide an engineering economic perspective on this issue. This book intends to provide a full-scope economic perspective on planning, implementation, operations, and maintenance of resilience enhancement measures in power infrastructure in the face of climate shocks and natural disasters. Our goal is to equip readers, including industry practitioners, policymakers, investors, and academic researchers, with the knowledge required to analyze power infrastructure resilience from an engineering economic lens.

Acknowledgments

Several chapters of this book were written based on valuable contributions made by the following co-authors:

Mohammad Shahidehpour, PhD
Suresh Khator, PhD
Zhu Han, PhD
Saeed Nematshahi
Eylem Tekin, PhD
Matthew P. Thompson, PhD
Yu Wei, PhD
Erin Belval, PhD
Rozhin Eskandarpour, PhD
Kevin Ding, PhD
Valentine A. Emesih
Behrouz Sohrabi

1

Introduction

1.1 The Art of Climate Resilience

Climate change is one of the most complex issues of our time. Addressing climate change as one of the grand challenges of the 21st century involves a two-pronged approach, which includes: (i) *climate adaptation*, i.e., adjusting to actual or expected future climate scenarios; and (ii) *climate mitigation*, i.e., reducing and stabilizing the flow of heat-trapping greenhouse gas emissions in the atmosphere [1]. William D. Nordhaus, a co-recipient of the 2018 Nobel Memorial Prize in Economic Sciences, describes climate change as the ultimate challenge for economics [2], calling for collaborative efforts by all sectors of society, including government, business, and civil society.

Energy systems, including the electric power infrastructure, are at the epicenter of this critical challenge, necessitating multidisciplinary collaboration among research communities to understand, assess, predict, and proactively respond to this climate phenomenon. This collaboration may be viewed from at least two critical lenses. The first lens focuses on climate change's impacts on energy systems and their interdependent critical infrastructure systems coupled with other techno-socioeconomic systems. The second lens is centered around the impacts of energy systems on climate change and its cascading effects on the different areas of *planetary boundaries* within which humanity can continue to develop and thrive for generations to come [1]. The concept of planetary boundaries identifies nine processes that are critical for maintaining the stability and resilience of the Earth system as a whole. As shown in Figure 1.1, six out of nine planetary boundaries have already been breached, while simultaneous pressure within all boundaries is increasing. Transgression of these boundaries identified in Earth system models indicates that the anthropogenic impacts on the Earth system must be considered in a systemic context. According to these models, if the planetary boundaries are respected, then global environmental functions and life-support

The Economics of Climate Resilience in Power Infrastructure, First Edition. Ali Arabnya and Amin Khodaei.

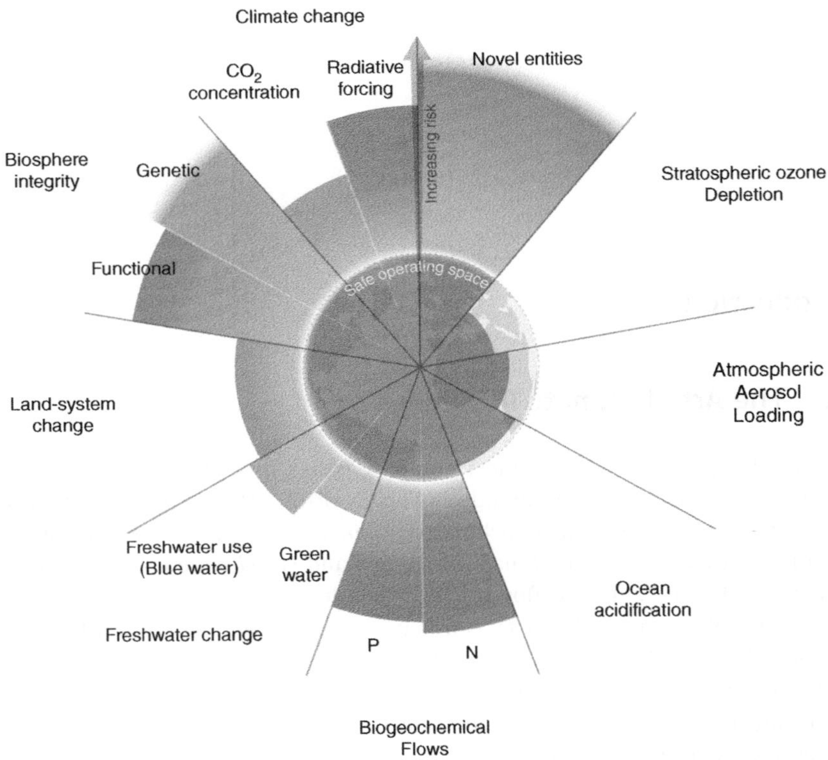

Figure 1.1 The 2023 update to the planetary boundaries. Source: Stockholm Resilience Centre, based on analysis in [3].

systems on Earth would be allowed to remain in their interglacial state—similar to those experienced over the past ~10,000 years—rather than transitioning into an uncertain state without prior analog in human history [3].

Both climate mitigation and climate adaptation efforts involve making decisions under *deep uncertainty* over time in dynamic interaction with the Earth system. Deep uncertainty exists when parties to a decision do not know or cannot develop consensus over (i) a system model that relates actions to outcomes; (ii) uncertainty of inputs to the model; and (iii) outcomes to be considered among various possible outcomes and their relative importance [4]. In other words, decision-making under deep uncertainty can involve situations where "we don't know what we don't know" about a system.

From a climate mitigation perspective, transitioning to net-zero emissions (NZE) energy systems by the middle of this century is at the center of climate action in the power and energy industry. Achieving the NZE milestone by 2050 requires a whole-of-society (WoS) approach to ensure that a set of policies; technologies; and societal, political, and economic solutions are developed and work in sync to achieve realistic decarbonization targets under economic and time constraints. There is a wide range of transition pathways that can be adopted to achieve decarbonization targets. The adopted pathway, however, will directly impact the decarbonization timeline and economic variables for stakeholders in the process. Reaching these targets without considering the short-, medium-, and long-term economic costs and benefits of reaching and maintaining these targets can result in serious impediments to the sustainability of NZE energy systems. Therefore, sustainability of the decarbonization pathways can play a critical role in streamlining their success. The National Environmental Policy Act (NEPA) of 1969 in the United States [5] defines sustainability as "to create and maintain conditions under which humans and nature can exist in productive harmony, that permit fulfilling the social, economic, and other requirements of present and future generations." The United Nations' Sustainable Development Goals (SDGs) provide a framework for peace and shared prosperity for people and planet Earth, now and into the future. At its heart are 17 SDGs with SDG 7 focusing on affordable and clean energy and SDG 13 addressing climate action involving mitigation and adaptation efforts [6]. To achieve this shared goal of sustainability, we need to build a global economy that meets the needs of our society without violating the planetary boundaries of the environment, as illustrated in Figure 1.2.

From a climate adaptation perspective, climate resilience enhancement measures are at the core of addressing this challenge for the power and energy industry. Climate change poses an increasing risk to our critical infrastructure and requires significant investment for upgrades, mitigation, and adaptation. Power infrastructure, in particular, is expected to be highly stressed by extreme events. Physical, operational, and financial losses caused by climate-related events such as wildfires, heatwaves, floods, and hurricanes can reach hundreds of billions of dollars, threatening the financial health and solvency of the electric utilities and local governments. In addition, these losses can significantly hinder the ability of utilities and infrastructure owners to recover from these events—which can lead to dire consequences for shareholders, ratepayers, and policymakers. This

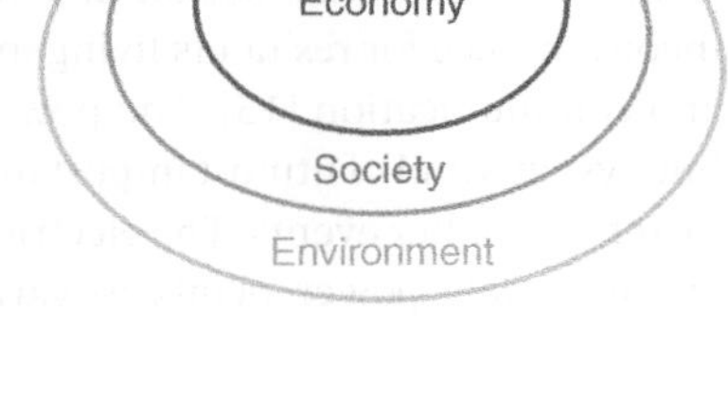

Figure 1.2 Sustainable economic development framework.

requires developing a wide range of innovative engineering solutions, financial mechanisms, and policy incentives to achieve the financial, operational, and physical resilience levels expected from a 21st-century power infrastructure.

It is further worth mentioning that while climate mitigation strategies (e.g., emission reduction) require global cooperation and local action, climate adaptation strategies (e.g., resilience enhancement measures) primarily require local solutions that form our first line of defense against climate extremes. A wide range of technological, economic, political, and community-based solutions must work together to achieve the desired level of resilience in the face of an ever-changing uncertainty due to climate shocks and natural disasters. This process is more of an art than a science!

1.2 Climate Risk in Power Infrastructure

Climate change poses an increasing risk to critical interdependent infrastructure systems, requiring significant investment to mitigate the risk in electric power infrastructure and the communities that are served. There are several studies in the literature indicating that climate change has increased the frequency and intensity of disasters in recent years [4–10]. The Fifth Assessment Report [11] by the Intergovernmental Panel on Climate Change (IPCC) concludes that "It is virtually certain that there will be more frequent hot and fewer cold temperature extremes over most land areas on daily and seasonal time scales, as global mean surface temperature increases." As a result, power systems are expected to be highly stressed by climate shocks due to ever-increasing power demand, reduced operational capacity during extreme climate events, and increased probability of system failures that can result in significant power outages [12]. As the risk landscape for climate shocks becomes more complex, it exposes the natural and built environments, including power infrastructure, to an unprecedented level of vulnerability.

Over half a century after the publication of one of the earliest studies on efficient response to climate-related events [13]—motivated by Hurricane Carla that slammed into the Gulf Coast and moved onward into the United States and Canada—the issue of efficient response to climate shocks and natural disasters still seems to remain in its immature stage [14]. Climate shocks can result in significant economic, social, and physical disruptions, and cause considerable inconvenience for residents living in disaster areas due to loss of electricity, water, and communication [15]. The power infrastructure, one of the most critical lifeline systems and of utmost importance to our daily lives, is increasingly exposed to extreme weather events. The electric power grid transfers the electricity generated by large-scale power plants to various industrial, commercial, and residential customers via transmission and distribution networks; hence, it is dispersed over a vast geographical area which increases its exposure to extreme events. For instance, more than 2.8 million customers experienced a power outage in Houston, Texas area after Hurricane Ike hit the United States in 2008, which lasted from a few days to several weeks. Damages to the coastal and inland areas were estimated at $29.5 billion [15].

The structure of the decision-making process before, during, and after climate shocks is an integral part of a disaster risk reduction strategy. Decentralized versus centralized decision-making is an age-old question. Each of these decision structures can outperform its alternative depending on the decision context and the decision maker's goals and performance metrics for the decisions made. When *responsiveness* through immediacy (i.e., taking the right action as quickly as possible in response to threats and opportunities) is the primary goal, decentralized decision-making can outperform its centralized alternative. However, when other decision metrics, such as *reliability* through compliance, *efficiency* through syndication, or *perennity* through detachment, are the primary goals in decision-making, centralized decision-making can be more advantageous than its alternative [16].

In the context of disaster risk management in power infrastructure, given the importance of responsiveness in the decision process, decentralized decision-making generally has more advantages over centralized decision processes [17]. Therefore, in the face of climate shocks and natural disasters, special attention should be given to a decentralized approach to disaster response and recovery resource allocation in power grids. The notion of decentralization in this context can be seen from at least two lenses: (i) decentralization of tasks and activities related to disaster response in power grids; and (ii) decentralization of organizational structure due to deregulation of the power industry, where generation, transmission, and distribution companies operate as independent entities. In both structures, however, coordination of tasks and activities with a central unit such as an Independent System Operator (ISO) or a command and control (C&C) center is required to improve the reliability and efficiency of these decisions.

Identification of climate-related hazard risk involves specifying hazards of concern based on stakeholders' prioritized list of concerns. The hazard prioritization should be based on (i) the likelihood of the hazard occurrence; (ii) the likelihood of severe consequences should the hazard be realized; (iii) strategic priorities; and (iv) the availability of resources and budget constraints. Hazard risk identification also includes the development of hazard scenarios and associated uncertainties. Development of hazard scenarios includes detailing the specific hazard conditions. The hazard scenario should include the expected location, time, and duration of the event, as well as other conditions required to adequately characterize the hazard and its impacts on the power system. Once the hazard risks are identified, measuring the exposure of the power grid assets to the identified risks is an essential element of enhancing resilience and managing the disaster risk. This includes a specific assessment of the scope and value of physical and virtual assets and systems that might face a destructive event under various hazard risk scenarios identified in the previous step. In addition to exposure to assets, specific customers and loads under exposure to any given hazard scenario must be determined. Determining the level of disruption requires a specific assessment of the expected level of damage or stress in exposed assets under various hazard scenarios. This involves an assessment of the vulnerability, i.e., the expected consequences to exposed assets when a destructive event occurs. Specifying the disruption level should not only indicate which assets are damaged or degraded, but should also specify how severely the assets are impaired, what the consequences are, and what steps are required to recover the overall system functionality to its pre-disaster conditions [18]. For instance, expected physical damage to grid assets and loss of power due to a heatwave event might include burnout of a high-voltage transformer, experiencing load interruption due to loss of this transformer and reduction of transmission capacity in a line, requiring several hours to bring the system back to its pre-disaster conditions, and resulting in a significant cost in repair expenses and loss of revenue.

Once the three elements of disaster risk management (i.e., the probability of an identified hazard, power system exposure, and power system vulnerability) are identified, a set of system models is required to collect the data and estimate the expected consequences. Data collection from historical events that characterize the intensity and duration of the disruption to the power system and the community is required to assess the power grid's resilience. Power utilities can maintain and improve the quality of data in their Outage Management System (OMS) to be used as a crucial input data set for a forward-looking resilience analysis process. System-level, data-driven predictive models capable of computing necessary power disruption estimates should be developed and regularly updated. These models should use risk data from previous steps (risk identification, risk exposure, and system vulnerability) as inputs to predict how the power system will be disrupted under various scenarios. For instance, expected physical damage due to a heatwave event can be used as input to a system model that relates them to energy not served (ENS) within the system over time. A set of system models may be required to capture various aspects of the system disruption, including power outage estimation models, restoration models, cost estimation models, and physical damage estimation models, among others. In addition, the interdependencies between models should be considered the system models as well [19]. For example, estimating the expected damage to the grid under various hurricane scenarios with different intensity, duration, and landfall trajectories requires a system model that uses three other system models, as follows:

$$ED = \sum_{S} Pr(S) \times EX(S) \times D(S) \tag{1.1}$$

where

ED is the expected damage to the system due to a climate-related event
S represents different scenarios
Pr(*S*) is a system model that gives the probability of a given scenario
EX(*S*) computes the exposure of the system under a given scenario
D(*S*) estimates the percentage of exposed assets that a loss will be materialized under a given scenario.

As another example, consider a restoration model that can determine the optimal restoration schedule of the power grid after a given natural disaster scenario. The schedule determined by such a model can inform systems models to assess system performance during the restoration and the required recovery time. The system models should be used to calculate the consequence estimates and resilience metrics defined in previous steps.

When uncertainty is involved in the analytical process, system models should be able to provide multiple probabilistic consequences and resilience metrics based on hazard scenarios, exposures, and vulnerability of the system obtained in previous steps. This will enable a risk-based decision-making process on resilience enhancement measures and prioritization of the mitigation options. Due to the probabilistic nature of this information and the uncertainty associated with the outcomes of these decisions, a specific risk metric based on both measures of central tendencies (e.g., mean) and dispersions (e.g., variance) can be used to describe the potential extent of these extreme events. When system models incorporate uncertainty in consequence estimates and resilience metrics, it is necessary to specify the statistical format of the metric, for example, the mean and variance of consequence estimates, the upper and lower bounds of consequence estimates, and the exceedance probability curve that describes the probability that certain loss levels will be exceeded. These risk metrics should be reflective of the stakeholders' risk perspectives [18, 19]. As an example, Table 1.1 describes four risk metrics that can statistically measure the potential damage from a climate-related scenario in power infrastructure.

There are various approaches for identifying the impact of events on the level of disruption. For instance, a model-based approach and analysis will allow the electric utility to plan and differentiate between two planning options. It will aim to provide a tool to inform the most efficient investment to mitigate a climate-related hazard. The value of impacted customers, outage duration, level of disruption, risk spend efficiency, and other metrics can be evaluated for several different investment prioritization strategies [18].

Table 1.1 Select risk metrics for measurement of damage due to climate-related events [19, 20].

Risk metric	Description
Value-at-risk (VaR)	A measure of risk that involves determining the worst loss expected over a target horizon within a given confidence interval.
Conditional value-at-risk (CVaR)—also known as expected shortfall	A risk assessment measure that quantifies the amount of tail risk, which is derived by taking a weighted average of the "extreme" losses in the tail of the distribution of loss, beyond the VaR cutoff point.
Probable maximum loss (PML)	The maximum loss expected from a disaster event, expressed in dollars or as a percentage of the total value of the asset.
Maximum foreseeable loss (MFL)	The worst loss that is likely to occur because of a single disaster event.

1.3 Resilience Metrics in Power Systems

A resilient power grid that ensures an uninterrupted electricity supply to citizens and interdependent lifeline systems is the keystone of modern society. The power grid continues to evolve as the needs and demands of society continue to change, and grid operators are faced with an emerging risk landscape. The changing power demand, increased reliance on renewable and distributed energy resources, and growth in penetration of smart technologies introduce additional layers of complexities to grid operations. Furthermore, the increasing frequency of natural disasters [21], climate change [22], manmade disasters (e.g., 2014 physical attack on PG&E's substation [23]), and cyber-attacks (e.g., malware BlackEnergy) pose a significant level of compounded risk to the uninterrupted operations of the modern power grid.

The complexity and interdependency of power infrastructure systems, either in urban or rural settings, can increase with time and are vulnerable to disasters. Developing mitigation strategies that outflank the process of risk transference of mega-disasters is the key to successfully managing disasters. In this context, *resilience* refers to the capacity to "bounce back" to a pre-disaster condition [24]. Based on the definition from [25], "local resiliency with regard to disasters means that a locale is able to withstand an extreme natural event without suffering devastating losses, damage, diminished productivity, or quality of life and without a large amount of assistance from outside the community."

Commonly used reliability metrics—such as System Average Interruption Duration Index (SAIDI), System Average Interruption Frequency Index (SAIFI), Customer Average Interruption Duration Index (CAIDI), and Customer Average Interruption Frequency Index (CAIFI), among others—are highly effective in measuring the reliability of the system and ensuring grid preparedness for disruptions and expected outages that occur under relatively normal conditions. However, due to the changing climate risk landscape, there is a broad consensus across the board that these metrics are inadequate to effectively address many of the emerging risk exposures in the power grid [26]. This is due to the fact that commonly used reliability metrics typically do not include outage information when low-probability, high-consequence disruptive events occur. In this emerging risk landscape, reliability metrics, which are built on historical data, may not be appropriate for capturing the complex risk dynamics as emerging risks can significantly deviate from their historical precedents. Similarly, security measures do not apply to grid resilience. While power grid security is primarily concerned with preventing a disruptive event from occurring in the first place, grid resilience is primarily concerned with ensuring the continuity of the service by the power grid to the communities that rely upon them in the face of disruptive events. Even though it may not be feasible to guarantee the continuity of operations at nominal or pre-disruption levels, ensuring a reduced service level can minimize the hazardous impacts of these events on communities. Therefore, a new set of metrics is required to address the resilience requirements of the evolving power grid.

Enhancing and maintaining power grid resilience requires not only a well-thought-out strategic vision and a well-executed action plan but also an effective allocation of capital and resilience investment prioritization under budget constraints to ensure the desired return on investment and the achievement of resilience goals. Specifying the resilience goals is the first step in the resilience optimization framework, laying the foundation for all steps in the process. To set resilience goals, senior management in electric utilities should form a cross-functional team of experts from key stakeholders to determine the acceptable (or unacceptable) impact of a given undesired event under various circumstances. Discussion during this phase is required to decide whether the resilience goals should be set based on (i) previous historical events, or (ii) aspirational resilience goals set by senior management. It is important to quantify the resilience goals in terms of acceptable duration, amount of lost load, and recovery cost for each load per each specific outage event. That quantification should be as specific as possible, including detailed information on the geographical boundary (e.g., urban, rural), load type (residential, commercial, industrial, or critical loads such as hospitals, police stations), hazard type (e.g., hurricane, wildfire), hazard intensity (e.g., a category 4 hurricane), and the time period after an event strike (e.g., the first hour, the first week). Once the resilience goals are defined, determining the consequence categories that serve as the basis for resilience metrics is the second step in the process. The consequence categories should be reflective of the resilience goals. In some instances, the consequence estimation and resilience metrics may focus on the impacts directly realized by the power utility, such as power not delivered, loss of revenue, and cost of recovery, among others. However, in some instances, direct impacts are only part of the resilience assessment process. Power and energy systems provide energy not just for generation or distribution but for some larger community benefit (e.g., transportation, healthcare, manufacturing, economic gain). Resilience analyses that aim to include a broader community perspective may convert power disruption estimates into community consequence estimates (e.g., number of emergency service assets affected, business interruption costs, impact on gross regional product, etc.). All the consequence categories are measured for the defined system specifications and therefore may be measured across spatial (geographical) and temporal (duration) dimensions. Data availability may also affect the selection of consequence categories. Resilience analyses are not restricted to a single consequence category for developing resilience metrics. Instead, a scenario-based approach using multiple consequence categories can be beneficial for representing various stakeholder perspectives [18, 19].

1.4 Economic Impacts of Climate Shocks and Natural Disasters

When a climate-related extreme event strikes, it affects not only physical assets including power infrastructure but also businesses and their ability to contribute to the local economy. These multidimensional, second-order effects are linked to both the physical characteristics of these extreme events (e.g., hurricane category, flood levels, heatwave intensity and duration) as well as the socioeconomic characteristics of the affected communities. The socioeconomic disparities of the affected communities shape the severity of shocks on local economies as well as the duration of subsequent recovery process. This is due to the fact that disaster consequences can remain long after they occur. A recent analysis of socioeconomic impacts of natural disasters by the author indicates that the impacts on the economic wellbeing of the affected communities can be far more than the cost of recovering the physical damages to the infrastructure. It also indicates that the recovery process not only is a function of the extent of physical damages due to disasters but also the economic structure and the socioeconomic resilience of the affected communities. The study shows that the value of lost income in the affected communities is far greater than the cost of damages to the infrastructure assets due to natural disasters [27].

The United States alone has sustained over 390 climate-related disasters from 1980 to mid-2024 where cost of damages due to each of these events was at least $1 billion (adjusted for inflation as of 2024). The total cost of damage due to these events reached over $2.755 trillion ($61.3 billion per year), as shown in Figure 1.3 [28]. The U.S. Department of Energy estimates that the annual economic cost of power outages in the United States alone is over $150 billion (adjusted for inflation as of 2015) [29]. Among them, about 83% of reported major power outages between 2000 and 2021 in the United States are estimated to be due to weather-related extreme events. Moreover, the average number of weather-related power outages per year is estimated to have increased by about 80% since 2011. From 2000 to 2021, there were over 1500 weather-related power outages leading to an estimated seven hours of outage per year for an average home or business in the United States [30].

Despite the increasing traction of climate economics in the power and energy community, there is still a lack of concerted efforts to address this gap using a multidisciplinary, multisectoral approach. Without considering the economic viability, business imperatives, cost of natural capital, political realities, and most importantly the socioeconomic resilience level of the communities, any engineering solutions for climate mitigation and adaptation will remain just an interesting idea that may not prove to be sustainable in the long run.

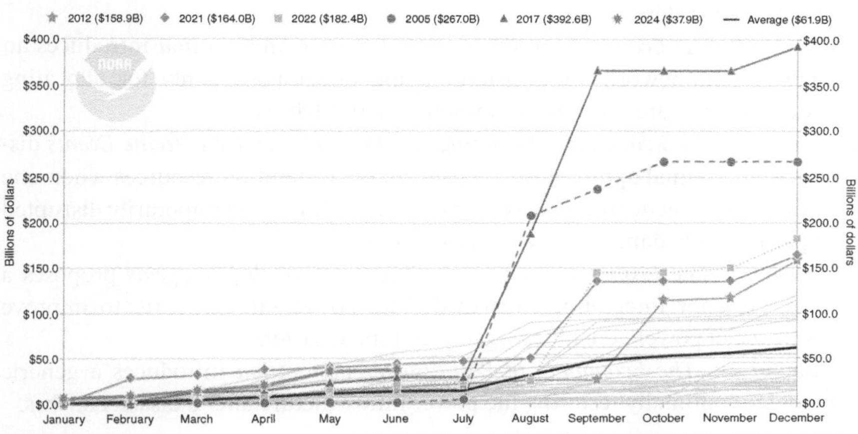

Figure 1.3 1980–2024 year-to-year cost of billion-dollar climate-related disasters in the United States [28].

1.5 Overview

This book discusses the engineering economics of climate resilience in electric power infrastructure by covering the economic decision-making processes involved in the design, planning, operation, and restoration of the grid in the face of climate shocks and natural disasters. The remainder of the book is organized as follows:

- Chapter 2: *The Economics of Financial Resilience Against Natural Disasters* presents an analytical framework for a risk transfer mechanism for power infrastructure.
- Chapter 3: *The Economics of Resilience-centered Asset Management* presents an integrated maintenance strategy for critical components of power infrastructure incorporating the risk of climate shocks and long-term climatological patterns.
- Chapter 4: *The Economics of Capacity Planning for Climate Shocks* introduces a stress testing method to measure extreme temperature impacts on power systems' transmission capacity.
- Chapter 5: *The Economics of Resilience Planning in Power Infrastructure Expansion* introduces a long-term capacity planning model considering future disruptions in the generation and transmission facilities due to climate shocks and natural disasters.
- Chapter 6: *The Economics of Operational Risk Management in the Face of Natural Disasters* presents a risk-rating framework for power infrastructure in the face of natural disasters.
- Chapter 7: *The Economics of Pre-disaster Resource Mobilization* introduces an analytical framework for decision-making under uncertainty for allocating restoration resources before an imminent natural disaster.
- Chapter 8: *The Economics of Contingency Planning Under Extreme Events* discusses an optimal approach to allocate energy generation resources when one or more of the generation units or transmission lines are temporarily disrupted or permanently damaged due to natural disasters.
- Chapter 9: *The Economics of Decentralization through Microgrids* proposes a framework for integrating microgrids into power infrastructure to improve resilience of power generation against natural disasters.
- Chapter 10: *The Economics of Post-disaster Restoration* introduces a generic model for optimally restoring the power infrastructure after a disaster strikes.

Each chapter of this book is designed to stand on its own and has its own nomenclature, bibliography, and acronyms.

References

1 A. Arabnya *et al.*, "Guest editorial: Climate change mitigation and adaptation in power and energy systems," *Int. J. Electr. Power Energy Syst.*, vol. 161, p. 110152, 2024.

2 W. Nordhaus, "Climate change: The ultimate challenge for economics," *Am. Econ. Rev.*, vol. 109, no. 6, pp. 1991–2014, 2019.

3 K. Richardson *et al.*, "Earth beyond six of nine planetary boundaries," *Sci. Adv.*, vol. 9, no. 37, pp. 1–16, 2023.

4 The society for decision making under deep uncertainty (DMDU). Accessed: Apr. 2024. Available: https://www.deepuncertainty.org/.

5 The National Environmental Policy Act (NEPA), 42 U.S.C. §§ 4321 et seq., 1969.

6 UN sustainable development goals. Accessed: Jun 2024. Available: https://sdgs.un.org/goals.

7 V. H. Dale *et al.*, "Climate change and forest disturbances: climate change can affect forests by altering the frequency, intensity, duration, and timing of fire, drought, introduced species, insect and pathogen outbreaks, hurricanes, windstorms, ice storms, or landslides," *Bioscience*, vol. 51, no. 9, pp. 723–734, 2001.

8 M. K. Van Aalst, "The impacts of climate change on the risk of natural disasters," *Disasters*, vol. 30, no. 1, pp. 5–18, 2006.

9 S. Banholzer, J. Kossin, and S. Donner, "The impact of climate change on natural disasters," in *Reducing Disaster: Early Warning Systems for Climate Change*. Dordrecht, The Netherlands: Springer, 2014, pp. 21–49.

10 M. Goss *et al.*, "Climate change is increasing the likelihood of extreme autumn wildfire conditions across California," *Environ. Res. Lett.*, vol. 15, no. 9, pp. 094016, 2020.

11 National Academies of Sciences, Engineering, and Medicine, *Enhancing the Resilience of the Nation's Electricity System*. Washington, DC, USA: National Academies Press, 2017.

12 IPCC, *Climate Change 2014 Synthesis Report: Contribution of Working Groups I, II and III to the Fifth Assessment Report of the Intergovernmental Panel on Climate Change*. Geneva, Switzerland: IPCC, 2014.

13 H. E. Moore, F. L. Bate, M. V. Layman, and V. J. Parenton, *Before the Wind. A Study to the Response to Hurricane Carla*. Washington, DC, USA: National Academy of Science-National Research Council, 1963.

14 A. Arab, A. Khodaei, S. K. Khator, and Z. Han. Electric power grid restoration considering disaster economics. *IEEE Access*, vol. 4, pp. 639–649, 2016.

15 R. Berg and National Hurricane Center, "Hurricane Ike: November 5–9, 2008," United States National Oceanic and Atmospheric Administration's National Weather Service, Tropical Cyclone Rep., Jan. 23, 2009.
16 H. Vantrappen and F. Wirtz, "When to decentralize decision making, and when not to," *Harv. Bus. Rev.*, vol. 26, 2017.
17 Z. Scott and M. Tarazona, "Study on disaster risk reduction, decentralization and political economy." Global Assessment report on disaster risk reduction, United Nations, 2011.
18 F. Petit, V. Vargas, J. Kavicky, M. Kintner-Meyer, and J. Eto. "Grid modernization: Metrics analysis (GMLC1.1)—resilience (Reference Document, Volume 3)," No. PNNL-28567, Richland, WA, USA: Pacific Northwest National Lab., 2020.
19 E. D. Vugrin, A. R. Castillo, and C. A. Silva-Monroy, "Resilience metrics for the electric power system: A performance-based approach," No. SAND2017-1493. Albuquerque, NM, USA: Sandia National Lab., 2017.
20 International Risk Management Institute (IRMI) Glossary. Accessed: May 1, 2024. Available: https://www.irmi.com/glossary
21 Swiss Re, "Natural catastrophes and man-made disasters in 2015: Asia suffers substantial losses," 2016. Accessed: Sep. 30, 2016. Available: http://media.swissre.com/documents/sigma1_2016_en.pdf
22 U.S. Department of Energy, "U.S. energy sector vulnerabilities to climate change and extreme weather," 2013. Accessed: Sep. 30, 2016. Available: http://energy.gov/sites/prod/files/2013/07/f2/20130716-Energy%20Sector%20Vulnerabilities%20Report.pdf
23 R. Smith, "Assault on California power station raises alarm on potential for terrorism," 2014. Accessed: Sep. 30, 2016. Available: http://www.wsj.com/articles/SB10001424052702304851104579359141941621778
24 B. Obama, "Presidential policy directive 21: Critical infrastructure security and resilience," Washington, DC, USA, 2013.
25 D. Mileti, Ed., *Disasters by Design: A Reassessment of Natural Hazards in the United States.*Washington, DC, USA: Joseph Henry Press, 1999.
26 NARUC (National Association of Regulatory Utility Commissioners), "Resilience in regulated utilities, 2013," Accessed: Sep. 30, 2016. Available: https://pubs.naruc.org/pub/536F07E4-2354-D714-5153-7A80198A436D
27 Y. Kerblat, A. Arab, B. J. Walsh, A. L. Simpson, and S. Hallegatte, "Overlooked: Examining the impact of disasters and climate shocks on poverty in the Europe and Central Asia region," World Bank, GFDRR, 2021.

28 NOAA, “National Centers for Environmental Information (NCEI) U.S. Billion-Dollar Weather and Climate Disasters,” Accessed: Jul. 14, 2024. Available: https://www.ncei.noaa.gov/access/billions/, doi: 10.25921/stkw-7w73.

29 H. Messera, “Distributed generation: Cleaner, cheaper, stronger-industrial efficiency in the changing utility landscape,” Pew Charitable Trusts, 2015.

30 U.S. Department of Energy, “Keeping the lights on in our neighborhoods during power outages,”2023 Accessed: Jul 12, 2024. Available: https://www.energy.gov/gdo/articles/keeping-lights-our-neighborhoods-during-power-outages

2

The Economics of Financial Resilience Against Natural Disasters

2.1 Principles of Risk Management

Risk management is becoming increasingly important for power utilities and infrastructure owners. While the practice of risk management is well-established in general, it is still evolving in the power and utilities industry to address the distinctive needs of this industry in the face of climate change and natural disasters.

2.1.1 A Three-Lines-of-Defense Framework

The defense-in-depth (or deep defense) is a paradigm that has its origins in military strategy, which relies on multiple lines of defense rather than a single frontline. The strategy is focused on creating redundant barriers to impede the advancement of an attacker or a threat to lose momentum over time, creating an opportunity window to respond with backup plans to contain the threat [1]. Outside of military science, this concept has gained significant traction in financial crisis management, enterprise risk management [2], cybersecurity [1], vaccine development [3], and nuclear power plant safety [4], among others. In the context of wildfires, a military tactic that consists of Primary Alternate, Contingency, and Emergency (PACE) plans has been adopted for large fire suppression in the Wildland Urban Interface (WUI). This approach includes building multiple layers of firelines between an active fire front and key communities or infrastructures to be protected during suppression operations, especially when weather conditions are unfavorable [5].

With that background, we define a three-lines-of-defense framework for managing the risk of wildfires in electric power infrastructure. The first line of defense focuses on strategies to prevent wildfires from occurring in the first place. Recognizing that the first line of defense may not always hold, the second line of defense is focused on mitigation strategies and proactive response to minimize

Figure 2.1 Three lines of defense framework for wildfire risk management in electric power infrastructure.

hazardous impacts of wildfires on the power system and its surrounding natural and built environment, should a wildfire spark. Finally, if a wildfire sparks and spreads despite all the defensive measures in the first two lines, we need a third line of defense that is focused on resilience-building measures and recovery preparedness so the system can bounce back to its pre-wildfire condition as quickly as possible without suffering devastating losses. Each of these lines plays a distinct role within the utility's wildfire risk governance structure. Figure 2.1 illustrates the proposed framework and its subcomponents. The first line of defense includes predicting wildfires, timely detection of the fire-inducing faults in the grid, early detection of wildfires, effective asset management of grid infrastructure, and effective vegetation management to prevent wildfires. The second line of defense includes the deployment of wildfire simulators to enable proactive responses, preemptive de-energization of the grid as a last resort option, effective wildfire suppression tactics, real-time wildfire monitoring and tracking capabilities, and effective emergency management of the grid operations during wildfires. Finally, the third line of defense includes efficient recovery logistics, energy contingency plans for the post-disaster supply of energy, comprehensive disaster risk financing mechanisms, and obtaining buy-in from stakeholders and communities affected by the event. The proposed framework is not meant to be comprehensive but is a lens we use in the following sections to review the most relevant thematic research work and identify the gaps in the literature that can potentially affect the best risk management practices in the power and utilities industry. In addition, the proposed framework can be used as a roadmap for utilities to develop their wildfire risk management strategy.

2.1.2 Operational and Financial Resilience

The economic and financial losses caused by climate-related events such as wildfires, heatwaves, and hurricanes can reach several billion dollars, threatening the financial health and solvency of the utilities. Climate change poses an increasing risk to our critical infrastructure and requires significant investment for upgrades, mitigation, and adaptation. This has attracted significant attention from the federal government, including the most recent U.S. Infrastructure bill, which includes US$550 billion in new federal investments in America's infrastructure over a period of five years. This bill will partly focus on climate change mitigation, resilience, equity, and safety. The electric power grid is a critical infrastructure system that is considerably impacted by climate-induced extreme events. In 2014, the Intergovernmental Panel on Climate Change (IPCC) concluded that more frequent hot temperature extremes will occur over most land areas as global mean surface temperature increases [6]. According to National Geographic, "an explosion in the frequency and extent of wildfires worldwide is hindering recovery even in ecosystems that rely on natural blazes to survive [7]." Power grids are expected to be highly stressed by these events, combined with ever-increasing power demand, reduced operational capacity, and increased probability of system failures resulting in significant power outages [8]. In addition, these events can significantly hinder the ability of utilities to recover, which can lead to dire consequences for shareholders, ratepayers, and policymakers [9].

Developing and implementing a tailored disaster risk financing strategy will increase the ability of impacted parties to respond more quickly and resiliently to disasters [10]. A recent report by the World Bank Group indicates that integrated operational preparedness and financial preparedness—as two pillars of resilient critical infrastructure systems—are crucial in managing the climate risk in power infrastructure [11]. Proactive operational planning, availability of financial resources, and collaboration with stakeholders are integral parts of an effective strategy to increase the resilience of the power sector against climate risk.

Operational resilience against climate risk includes predetermined plans to energize the critical services in the short term and restore the power grid under the system operator's or relevant utilities' purview. These plans not only need a local research phase to obtain protocols to tackle the climate disaster but also need an implementation phase using experts and organized equipment. Also, it is imperative to simultaneously reinforce the key infrastructures, such as power lines and substations bearing the brunt of severe weather challenges. For example, in case of flood, elevating equipment would be beneficial, or in case of wildfire, integrating advanced insulation materials into infrastructure is recommended. In addition, developing an advanced monitoring and controlling system is also crucial for power system operators to make quick decisions in case of disruptive events. Many power utilities demonstrate exemplary resilience practices by integrating resilience projects into their rate base. For example, FPL took substantial steps to safeguard its assets from future impacts, including burying electric lines, reinforcing transmission poles, reducing pole spacing to mitigate wind damage, upgrading flood-resistant transformers, and collaborating with local stakeholders to safeguard critical services and infrastructure.

Financial resilience incorporates designing a mechanism that ensures quick access to sufficient financial resources to execute the prescheduled plans to restore power in case of failure caused by any kind of disaster. Therefore, it is necessary to carry out risk modeling to understand the potential cost of climate-related disruptions. Then, the grid operator can take advantage of risk transfer mechanisms and insurance options to cover the potential financial losses. However, little attention has been given to developing a financial protection strategy to address this challenge. For example, the 2018 Camp Fire alone in Northern California—sparked by Pacific Gas and Electric Company's (PG&E's) electrical infrastructure—made PG&E face a multibillion-dollar lawsuit [12]. Even with meticulously designed plans, funding plays the underlying role in the success of a resilience project. An appropriate mechanism for financial resilience incorporates mobilizing and delivering necessary funds. For instance, electric utilities in the United States are required to submit their emergency response plan to state public utility commissions for approval regularly. In the event of a power outage caused by climate risk, financial instruments will ensure access to funds for maintenance. However, timeliness is the highest priority when allocating funds to grid operators. Thus, it is crucial to ensure efficiency in requesting, approving, and disbursing the funds between the public and private sectors. Figure 2.2 illustrates the discussed efficient shock responsive system to protect the power grid from climate risk.

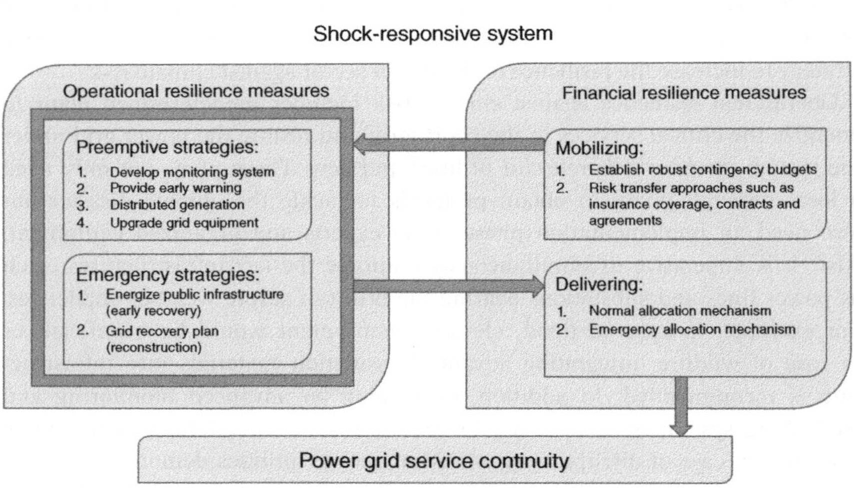

Figure 2.2 Grid resilience against climate shocks.

2.2 Climate Risk Financing Mechanisms

A wide range of disaster risk financing mechanisms can facilitate a utility's ability to access capital to cover the risk of climate-related events. These mechanisms are not necessarily mutually exclusive and should be bundled together to ensure the required coverage for various climate-related scenarios [9]. There are two approaches to cover the cost caused by climate disasters. The first strategy to meet the financial burden is to shift it to customer bills, and the second procedure is to finance the expenses by the utility. For some Investor-Owned Utilities (IOUs), the economic implications of climate-induced damages typically get transferred to ratepayers in the form of electricity bills. As an example of these significant losses, Xcel Energy, a public service company in Colorado, received \$2 billion in lawsuits for grassland wildfire in December 2021. For other IOUs, the financial impact of damages is shared by shareholders and ratepayers. As an example for this category, San Diego Gas & Electric (SDG&E) in California requested \$379 million to recover the 2007 southern California wildfire damage. However, the California Public Utilities Commission (CPUC) denied their request because it turned out the utility didn't meet the standards, and therefore, they were prohibited from recovering this cost through electricity bills. As another example of IOUs, after hurricanes in 2004 and 2005, FPL requested \$1.7 billion from ratepayers. After bargaining with the Florida Public Service Commission (FPSC), FPL collected funds from customers, around \$1 per month, from 2007 to August 2019, to cover over \$1.3 billion in losses. Our recent analysis [13] shows a wide gap in the literature and industry practices to provide innovative actuarial analysis and investigate the effectiveness of various disaster risk financing mechanisms in power grid infrastructure in the face of climate-related disasters. However, there are various disaster risk financing strategies that can be adopted by utilities, as follows:

- *Funded self-insurance* is an approach where a utility retains all or a fraction of the risk by setting aside adequately funded reserves to cover unexpected losses and restoration costs. For instance, a utility can create a climate reserve account that collects funding over time until it reaches a pre-determined account balance. This account should be designed to return the collected amount that exceeds the required climate reserve level to ratepayers. For example, on January 12, 2023, the CPUC approved a settlement agreement that allows PG&E, California's largest utility, to reduce the risk of wildfire liability by collecting hundreds of millions of dollars to self-insure. PG&E established a target of \$1 billion for a self-insurance account and set an initial \$250 million for 2023 as a starting point. A separate report from Washington State Electric Utilities clarifies that 8 utilities reported engaging in self-insurance in January 2023 while the remaining

23 did not [14]. While this mechanism can reduce the post-disaster financial burden to utilities and ratepayers, it will take time to build such reserves, and it may face regulatory challenges in some jurisdictions [9]. The superiority of self-insurance over other approaches is that the utility can access the account immediately, and there is no lag between the request time and the time the amount gets paid. On the other hand, the disadvantage of self-insurance is that if a climate disaster occurs that is costlier than expected, the account won't be able to cover it, which will impose financial stress. To overcome this issue, FPSC allowed utilities in Florida to impose surcharges on customers if the storm damage cost exceeds the reserve account. More details on economic theory and the optimal self-insurance against natural disasters can be found in [15].

- *Commercial insurance* is another mechanism that utilities have traditionally used to cover disaster risk. Under this mechanism, depending on the jurisdiction, the cost of insurance coverage can be distributed to ratepayers as climate premiums. Utility companies use different insurance coverages to handle specific risks. Commercial General Liability Insurance protects against injury, property damage, and libel/slander claims. Directors & Officers Insurance defends leaders from personal liability for wrongful acts. Commercial Auto Insurance covers the company's vehicles, while Commercial Property Insurance protects assets like buildings and equipment. Additional policies include Business Interruption Insurance for operating expenses during shutdowns, Extra Expense Insurance for new workspaces after perils, and Pollution Liability Insurance for cleanup costs. Boiler & Machinery Insurance guards against breakdowns, Cyber Liability & Privacy Insurance protects against cyber incidents, and Builders Risk Insurance covers construction projects. Wrap-Up Liability Insurance offers liability coverage during construction. These policies collectively form a strong risk management strategy for utility companies.
  However, the recent increase in the intensity and frequency of disruptive events have made this mechanism significantly more expensive for some utilities because of their increased risk exposure due to climate change [9]. The impact on the insurance industry has also become significant in recent years; private insurers in the United States paid over $50 billion in wildfire losses between 2017 and 2022, according to the American Property Casualty Insurance Association [16]. In this regard, insurers have increased their premiums or stopped issuing new policies altogether in some states (e.g., Allstate and State Farm in California). As another example, PacifiCorp has requested an accounting deferral from Oregon regulators, citing a fourfold increase in future commercial insurance premiums due to heightened wildfire risk in the West. Note that the commercial insurance may also cover any other climate-related events such as hurricanes, floods, and storms.

- *Catastrophe bonds* (also known as "CAT bonds") are a class of disaster risk financing mechanisms that can be used to transfer climate-related disaster risk from sponsors (utilities) to investors. The sponsors can form a Special Purpose Vehicle (SPV) financial entity to collect principal from investors and make premium payments in return. The premium spread depends on the expected loss (EL) for investors. If a triggering event with previously defined specifications in a defined timeframe occurs, the principal is given to the sponsor (utility in this case) so they can cover their loss; otherwise, the principal, along with return on investment, is returned to the investors [9]. CAT bonds are popular with investors since they are uncorrelated with other risks such as interest rate risk, equity market risk, and credit risk. Insurers and public agencies are turning to Wall Street to share the financial burden of rising natural disaster costs driven by climate change. The CAT bond market has surpassed $41 billion, doubling since 2013. Insurers like Amtrak, Google, and the World Bank use financial markets to limit damage bills from hurricanes, wildfires, and earthquakes. Fermat Capital Management Co. manages over $9 billion in CAT bonds, offering investors returns of over 10%, with limited correlation to traditional investments. Despite being risky, these bonds have gained popularity due to the growing frequency of climate-related disasters. The appeal lies in their ability to provide reinsurance protection, acting as a financial tool for insurers to transfer risks and attract capital. However, unsuccessful experiences also occur. For instance, investors who purchased a $200 million CAT bond from PG&E in August 2018 lost their entire stake when the devastating Camp Fire erupted three months later. While CAT bonds offer lucrative returns, they pose the risk of total principal loss in the event of a disaster. More details on CAT bonds can be found in [17]. Figure 2.3 demonstrates the major components and cash flows associated with CAT bonds.
- *Captives*, as an alternative insurance mechanism, are a class of insurance firms established by one or a group of companies to insure the owners (the parent companies). Captive insurance involves insureds establishing their own insurance companies and operating outside the commercial insurance marketplace to meet their risk financing objectives. One or a group of utilities can form a captive insurance entity that collects insurance premiums, issues climate policies, and pays out claims—like a traditional insurance company, but only to its utility parent companies. This mechanism is especially useful for covering uninsurable or difficult-to-insure risks due to increased exposure, such as post-2017 and 2018 wildfires in California. In addition, it can benefit utilities in hard insurance markets by reducing the risk transfer costs [9]. For example, CAT bond rates offered by investors skyrocketed after the abovementioned wildfires in California. In such cases, captive insurance is crucial for the companies. For example, On October 11, 2023, PG&E formed a new non-Rule

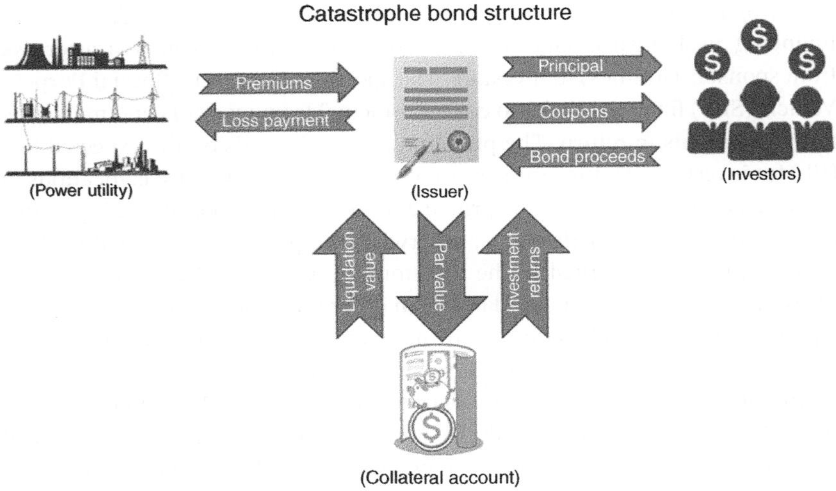

Figure 2.3 CAT bond components and cash flows.

II.B affiliate, Pacific Energy Risk Solutions LLC, to serve as a captive insurance company to transact insurance and manage PG&E's wildfire self-insurance program. Captive insurance models provide valuable tax-planning benefits, boosting the profitability of the parent company. By establishing itself as a legitimate insurance company, a captive allows the parent company to deduct the premium paid as an insurance expense. Moreover, captives offer gift and estate tax savings for shareholders and generate income tax savings for both the parent company and the captive. With its versatility, captive insurance efficiently covers a wide range of risks and is customized to meet the specific needs of the parent company. While traditionally established to address internal risks, there is a growing trend of captives underwriting third-party risks, including employee benefits, supply chains, and extended warranties. Essentially, captives can cover almost any risk handled by conventional insurance carriers. However, the permissible lines of coverage vary by state. For instance, Connecticut allows captives to insure against diverse losses such as life, health, commercial, and annuities, while Arizona restricts coverage for certain areas like workers' compensation or employee liability insurance. More details on optimal risk financing through captives can be found in [9].

- *Risk pooling* is a disaster risk financing mechanism for sharing risk among a group of participants. The participants combine their financial resources to cover losses whenever a pool participant experiences a loss that is covered in the pool. Risk pooling involves like-minded entities joining forces to share insurance costs. Instead of a conventional insurance approach, this model centers on distributing the burden of similar risks within the group. Annually, an actuary assesses the group's risk costs, which are then divided among members. Each member contributes their share to a central fund. In case of a claim, the central fund covers the costs. Due to the inherent uncertainty in estimating yearly risk costs, the risk pool conducts a financial review at year-end. If initial contributions were insufficient, members may need to pay extra to replenish the fund. Conversely, if excess funds exist, they can be utilized by the pool or distributed as dividends to members. This mechanism is beneficial when commercial insurance is impossible, difficult, or expensive to obtain, and there is an adequate number of participants that can form a stable risk-sharing pool [9]. For example, in wildfire-prone areas with increased exposure, such as California, where several utilities are servicing the area, risk pooling can play a crucial role in managing the financial risk of wildfires. The Association of Governmental Risk Pools (AGRiP) recommends that members of risk pools can generally experience a 15–25% reduction in insurance costs over time. Surplus funds can be returned to members or used for additional services. However, equitably implementing this approach is challenging due to varying climate risks in different areas, making it difficult to pool high-risk and low-risk areas together. More details on disaster risk financing through risk pooling can be found in [18].

- *Recovery bonds* are a class of post-disaster risk financing mechanisms (as opposed to pre-disaster financing mechanisms described earlier) that can be used to acquire capital to pay for unexpected damages from disasters. Depending on their jurisdiction, utilities can issue recovery bonds through various mechanisms so they can access a larger amount of capital to pay for reconstruction expenses and other liabilities. In return, the investors will receive a return on their investment based on coupon rates on the bonds. This mechanism has been used by utilities for financing hurricane damages [19] and has recently been legislated to cover the utilities' liabilities due to wildfire events in California [20]. Southern California Edison (SCE) sponsored $775 million in Senior Secured Recovery Bonds in 2023. This was built upon issuing two prior series: $338 million in Series 2021 Bonds and $533 million in Series 2022 Bonds. Notably, all these issuances comply with the Wildfire Financing Law, showcasing a consistent trend in using recovery bonds for power utilities. Similarly, following the 2004 hurricane season, Florida Power & Light (FPL) faced the aftermath of four named storms in 2005 (Dennis, Katrina, Rita, and Wilma). In response, FPL requested the FPSC to issue a financing order approving the issuance of storm-recovery bonds. Ultimately, the FPSC approved the issuance of storm-recovery bonds, amounting to up to $708 million. This example further highlights the utilization of recovery bonds in response to natural disasters and their approval through regulatory processes.

Climate risk is emerging as the Achilles' heel of the modern power grid, which threatens not only the continuity of operations but also the solvency of the electric utilities. The good news is that a variety of risk management strategies and market-based mechanisms are currently available to strengthen the financial resilience of utilities in the face of these events. However, there is no "silver bullet" to address this issue, and the utilities need to develop a holistic, multi-layer financial protection strategy for an optimal solution that meets their resilience goals and considers the economic constraints, risk landscape, and physical characteristics of their infrastructure.

2.3 A Financial Resilience Building Framework Using CAT Bonds

CAT bonds are financial instruments developed by insurance companies and reinsurers to transfer the risks associated with natural disasters to the capital markets. These bonds can offer either excess-of-loss (XOL) or binary (Bin) payouts, depending on the situation. The payout function for XOL (also known as layered) is defined as follows:

$$L(a, a+h] = \begin{cases} 0, & \text{if } L \leq a, \\ L-a, & \text{if } a < L \leq a+h, \\ h, & \text{if } L > a+h. \end{cases} \tag{2.1}$$

where the total loss is linked to a layer beginning at point a (attachment point) and extending to point $a+h$ (exhaustion point). According to this definition, if the wildfire loss is less than or equal to a, the utility cannot receive any funds from the collateral account. In the second scenario, the utility can access a portion of the principal if the total loss falls within the range greater than a and less than or equal to $a+h$. Finally, if the total loss surpasses the exhaustion point, the entire money is allocated to the utility to offset the wildfire losses. It is important to note that the total loss comprises both liability claims and power infrastructure losses.

In practice, for wildfire CAT bonds, the bandwidth represented by h is not equal to the principal amount. Therefore, we modified the payout function to (2.2) to accurately represent the relationship between loss and the total principal, as follows:

$$L(a, a+h] = \begin{cases} 0, & \text{if} L \leq a, \\ ((L-a)/h) * P, & \text{if} a < L \leq a+h, \\ P, & \text{if} L > a+h. \end{cases} \tag{2.2}$$

The payout function for binary form is:

$$L_{a,P}^{Bin} = \begin{cases} 0, & L \leq a, \\ P, & L > a. \end{cases} \tag{2.3}$$

where the payout will be either full or none. When the total loss is less than or equal to the threshold, i.e., a in (2.3), the utility will receive nothing; however, if the total loss exceeds the threshold, the utility will receive all the funds in the collateral account.

To calculate the EL, the Cumulative Distribution Function (CDF) for the loss is defined as follows:

$$F_L(l) = P(L \leq l). \tag{2.4}$$

CDF presents the probability of total loss being less than or equal to l. Based on this definition, the probability of total loss greater than l is the Decumulative Distribution Function (DDF) and is defined as follows:

$$S_L(l) = 1 - F_L(l) = P(L > l). \tag{2.5}$$

Given the existing DDF for stochastic variable L, it is straightforward to articulate the DDF for the loss layer within the interval $(a, a + h]$ as follows:

$$S_{L(a,a+h]}(y) = \begin{cases} S_L(a + y) = P(L > a + y), & \text{if } 0 \leq y < h, \\ 0, & \text{if } y \geq h. \end{cases} \tag{2.6}$$

This implies that if the loss amount falls within the specified layer, the corresponding DDF represents the probability of total loss exceeding a given threshold. Conversely, if the loss amount falls outside this layer, the DDF is set to 0.

Determining the anticipated loss of an insurance product involves computing the EL of the related layer. This is because the expected absolute loss for any arbitrary loss variable L (with a minimum value of 0) is provided by the below equation:

$$E(L) = \int_0^{\infty} S_L(l)dl, \tag{2.7}$$

The expected value of the absolute loss layer within the interval $(a, a + h]$ directly follows from:

$$E(L_{(a,a+h]}) = \int_0^{\infty} S_{L(a,a+h]}(y)dy = \int_0^{h} S_L(a + y)dy = \int_a^{a+h} S_L(l)dl. \tag{2.8}$$

which refers to the integration of the DDF from the point of attachment to the point of exhaustion. Subsequently, the EL of the layer can be defined as a function of the Probability of First Loss (PFL) and the Conditional Expected Loss Rate (CEL) as follows:

$$PFL = S_L(a) = P(L > a) \tag{2.9}$$

$$CEL = \frac{E(L_{(a,a+h]} \mid L > a)}{h} \tag{2.10}$$

$$EL = \frac{E\left(L_{(a,a+h]}\right)}{h} = P(L > a).\frac{E\left(L_{(a,a+h]}|L > a\right)}{h} = PFL.CEL \tag{2.11}$$

Based on a thorough comparison among several premium calculation models in [21], we use the linear premium principle, which is the recommended model. Therefore, the premium $P(l)$ for the interval $(a, a + h]$ is expressed as:

$$\rho(l) = EL + \Lambda = PFL \cdot CEL + \Lambda \tag{2.12}$$

The risk loading (Λ) ensures that investors receive fair compensation for assuming the risks linked to catastrophic events. This involves multiple factors, including investors' appetite for CAT bonds, legal fees, and SPV fees.

The precision of wildfire scenarios outlined in Section 2.4 will furnish us with a comprehensive assessment of total loss within the designated area of inquiry. Subsequently, this information will serve as the foundation for executing premium calculations.

2.4 Wildfire Simulation Modeling

We aim to quantify the risk of grid-ignited wildfires and propose an indemnity-based CAT bond for power utilities. Therefore, the initial step is to calculate an accurate pricing of the associated risk. As we consider both third-party liability and damages to the power infrastructure, it is necessary to integrate geographic and weather data as inputs to simulate wildfire spread in the study area. We use the U.S. Forest Service's open-source package, FARSITE [22], for wildfire simulation. The transmission grid's GIS map is needed to accurately place the ignition point on the power line. Subsequently, we will transmit the ignition point to FARSITE, and the information on the burned area will be overlaid onto the GIS map. This overlay will allow us to assess both liabilities and damages to the transmission system in the affected area. In the following, we provide a detailed demonstration of the data required for the modeling process and how we modeled them.

2.4.1 Geographic Data

The first dataset inputted into FARSITE consists of topographic data. The study region is located approximately 100 miles to the east of San Francisco Bay. It is situated within the latitudes of 37.6°–38.1° and longitudes of −120.7° to −120°, covering an area of around 800,000 acres, as illustrated in Figure 2.4.

Landscape data is extracted from FlamMap version 6.2, comprising eight layers: elevation, slope, aspect, fuel model, canopy cover, stand height, canopy base height, and canopy bulk density.

Figure 2.4 Geographic map for the study area.

2.4.2 Meteorological Data

The second set of data is the weather data. In this study, we include wind speed, wind direction, temperature, and humidity, as each significantly impacts the spread of wildfire. The historical data for the study region is retrieved from the National Solar Radiation Database (NSRDB) at hourly intervals [23].

2.4.3 Power Grid Data

An ignition point is required to initiate a wildfire scenario at this stage. Thus, we need to establish a geographic representation of the transmission system. We utilized the IEEE 30-bus test system along with a geospatial map provided in [24]. To better understand the transmission system's configuration, Figure 2.5 displays the power lines in black and the buses in red.

2.4.4 Ignition Points

Without loss of generality, our study focuses solely on wildfires ignited by power lines. Each ignition point represents a distinct wildfire scenario and is situated along the power lines. As shown in Figure 2.6, these ignition points are strategically distributed, maintaining an approximate distance of 2 km between them along the power lines. In total, 305 ignition points were established, which resulted in 305 distinct wildfire scenarios.

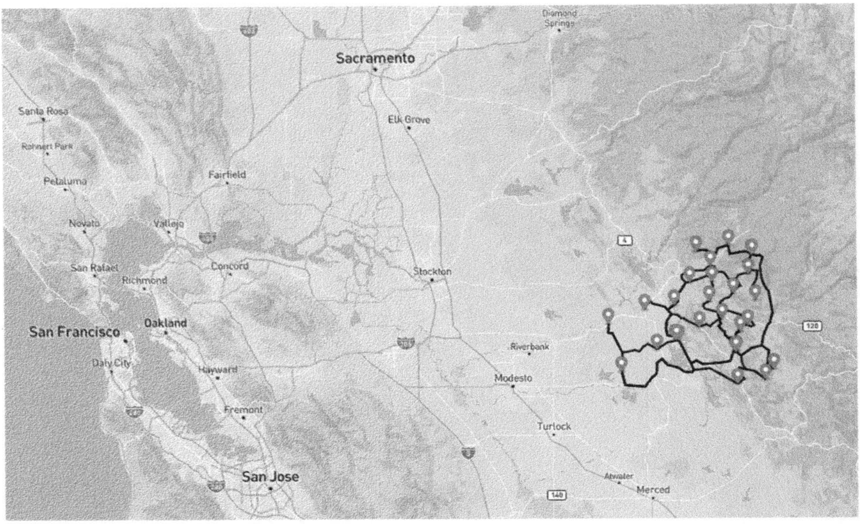

Figure 2.5 Geographic visualization of the IEEE 30-bus test system.

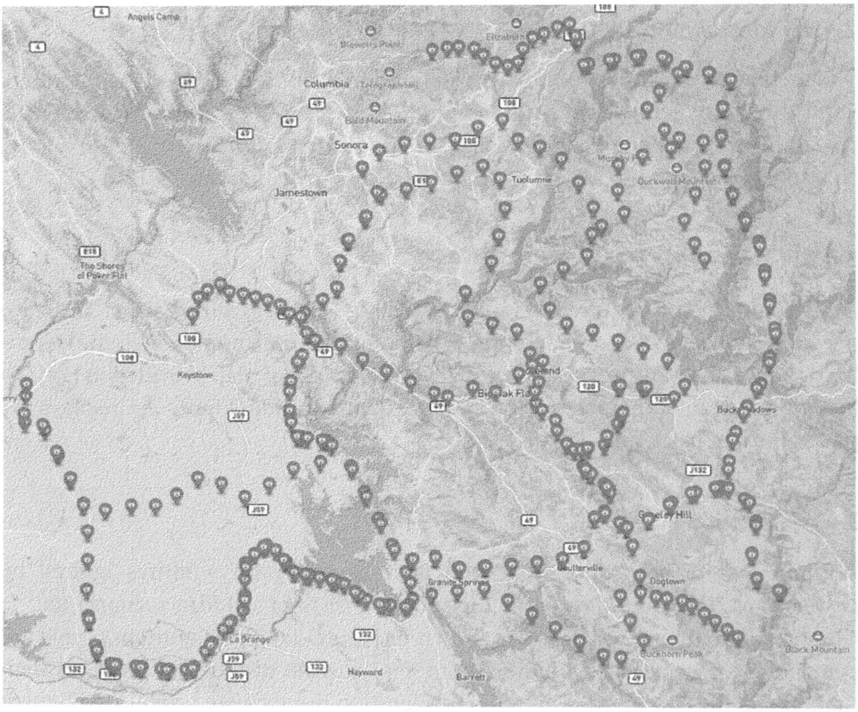

Figure 2.6 Spatial distribution of ignition points originating from power lines.

Now, we can analyze each distinct grid-ignited scenario to assess the associated risks in Section 5.

2.5 Case Studies

Having defined the 305 grid-ignited wildfire scenarios, assessing the total damage distribution, encompassing both liability and power line reconstruction costs is imperative to accurately price the associated risk.

2.5.1 Total Damages

By integrating geographical and meteorological data with specific ignition points, we obtain distinct burned areas for each scenario. Figure 2.7 compares each scenario's total burned area, measured in acres.

The distribution of burned areas in our small-scale power transmission lines demonstrates a notable dispersion from the average. Notably, this dispersion is even more pronounced in large-scale power transmission systems due to significant variations in geographic and weather data around distant ignition points. Consequently, the standard deviation in such cases tends to be higher, underscoring the importance of understanding and accounting for these diverse environmental factors.

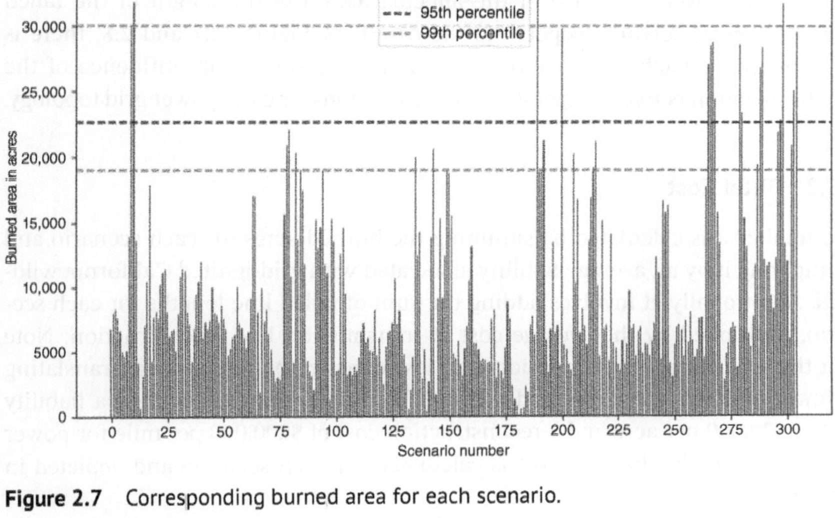

Figure 2.7 Corresponding burned area for each scenario.

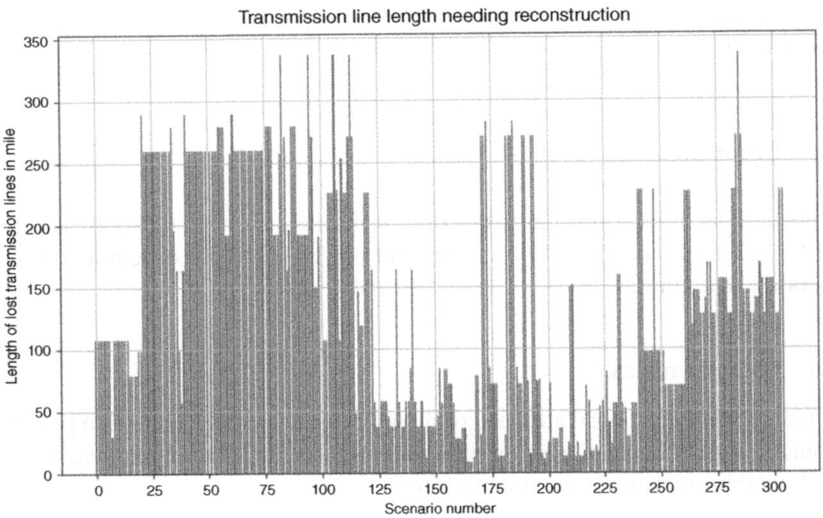

Figure 2.8 Comparison of failed lines needing reconstruction.

We can identify the affected lines by overlaying the burned area with the existing power lines. Subsequently, we calculate the length of power lines requiring reconstruction for each scenario. Figure 2.8 illustrates the specific lengths of power lines necessitating reconstruction in each scenario.

In certain extreme scenarios, restoring power transmission systems may necessitate replacing over 300 miles. At first glance, one might anticipate a high correlation between the size of the affected area and the length of the failed lines. However, delving deeper into the details of Figures 2.7 and 2.8, there is no correlation. Each significantly depends on the substantial influence of the synchronization between vegetation, fuel conditions, and the power grid topology.

2.5.2 Total Cost

The total cost is calculated by summing the burned acres for each scenario and multiplying it by an average liability associated with grid-ignited California wildfires. Additionally, it involves adding the sum of failed line lengths for each scenario, multiplied by the average cost of transmission line reconstruction. Note that the selected region is considered a low-density population area, translating to fewer affected structures and fatalities. In this study, we assumed a liability cost of \$20,000 per acre and a reconstruction cost of \$200,000 per mile for power lines. Accordingly, the total cost is calculated for each scenario and depicted in Figure 2.9.

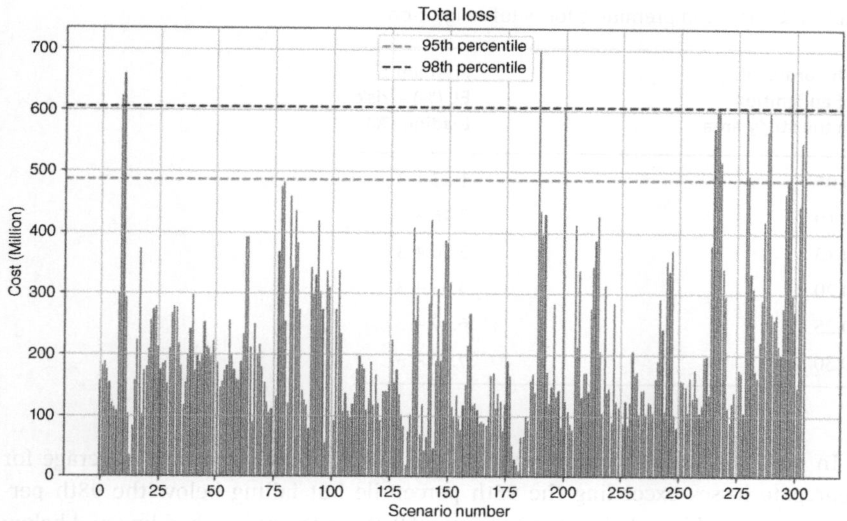

Figure 2.9 Comparison of total loss for each grid-ignited scenario.

The total loss ranges from $7 million to $700 million, with the highest total loss associated with scenario number 190. This scenario represents a wildfire starting from line 27 between buses 10 and 21.

2.5.3 Premium Calculation

Parametric insurance acts as a complementary component to conventional insurance models. The total bond yield, or coupon, transferred from the issuer to investors comprises two primary components: (i) a risk-free interest rate, typically based on the latest US Treasury notes, and (ii) a premium, encompassing the expected risk along with a risk loading factor. Figure 2.10 illustrates these components of the coupon.

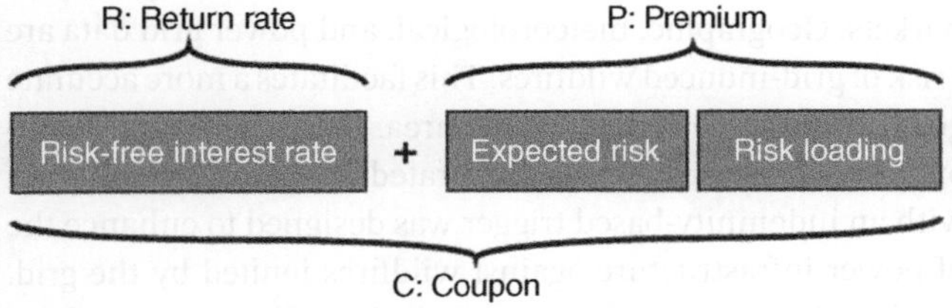

Figure 2.10 Composition of CAT bond yield components.

Table 2.1 Derived premiums for wildfire ignition.

The probability of an ignition in the study area	Premium: EL (%) + risk loading (%)
0.05	1.23 + 3
0.10	2.46 + 3
0.15	3.70 + 3
0.20	4.93 + 3
0.25	6.17 + 3
0.30	7.40 + 3

In our risk transfer strategy, we tailor the CAT bond to extend coverage for aggregate losses exceeding the 95th percentile but falling below the 98th percentile threshold, as delineated in Figure 2.9 above the red dashed line and below the black dashed line. Thus, the attachment point is set at $487 million, and the exhaustion point at $601 million. The preferred duration for CAT bonds often spans three to five years, offering a suitable timeframe for risk assessment and investor yield.

Applying a risk loading of 3%, Table 2.1 delineates the premium calculated according to the probability of ignition for grid-ignited wildfires within the designated study area.

2.6 Summary

In this chapter, we presented a strategic risk management framework for climate-related events with a special focus on wildfires. We presented various financial resilience building mechanisms that utilities can deploy in the face of climate shocks and natural disasters. We proposed a novel parametric insurance solution specifically designed for power utilities to transfer the risk of wildfire ignition to investors in capital markets. Geographic, meteorological, and power grid data are overlaid to model the risk of grid-induced wildfires. This facilitates a more accurate pricing of the damages within the designated study area. Both third-party liability and reconstruction costs of power lines are integrated into the calculation of losses. A CAT bond with an indemnity-based trigger was designed to enhance the financial resilience of power infrastructure against wildfires ignited by the grid. This tailored model, designed specifically for power utilities, offers a comprehensive perspective on the susceptibility of their power lines to trigger wildfires and subsequent financial consequences. Premium calculation is discussed separately to provide utilities with a higher level of information symmetry on their underlying risk and the expected cash flows involved in the CAT bond contracts.

References

1 The Department of Homeland Security, "Recommended practice: Improving industrial control system cybersecurity with defense-in-depth strategies," Washington, DC, USA, 2016.

2 The Institute of Internal Auditors, "IIA Position Paper: The three lines of defense in effective risk management and control," Altamonte Springs, FL, USA, 2013.

3 A.M. Sholukh *et al.*, "Defense-in-depth by mucosally administered anti-HIV dimeric IgA2 and systemic IgG1 mAbs: Complete protection of rhesus monkeys from mucosal SHIV challenge," *Vaccine*, vol. 33, no. 17, pp.2086–2095, 2015.

4 IAEA, "Defence-in-depth in nuclear safety. INSAG-10," A Report by the International Safety Advisory Group, Vienna, Austria, 1996.

5 "Tactical Tools for Wildland Fire," County Los Angeles Fire Dept., Los Angeles, CA, USA, Apr. 2016.

6 IPCC, *Climate Change 2014 Synthesis Report: Contribution of Working Groups I, II and III to the Fifth Assessment Report of the Intergovernmental Panel on Climate Change.* Geneva, Switzerland, IPCC, 2014.

7 J. Pickrell, "Wildfires have spread dramatically—and some forests may not recover," National Geographic, 2020.

8 National Academies of Sciences, Engineering, and Medicine, *Enhancing the Resilience of the Nation's Electricity System.* Washington, DC, USA: National Academies Press, 2017.

9 C. Kousky, K. Greig, and B. Lingle, *Financing Third Party Wildfire Damages: Options for California's Electric Utilities.* Philadelphia, PA, USA: Wharton Risk Management and Decision Processes Center, 2019.

10 Disaster Risk Financing and Insurance (DRFI) Program, *The World Bank.* Available: https://www.worldbank.org [Accessed: September 2020]

11 World Bank Group, "Financial Protection of Critical Infrastructure Services," 2021.

12 I. Penn, "PG&E says wildfire victims back settlement in bankruptcy," *The New York Times*, May 2020.

13 H. Jill, I. S. Wing, and M. Lisk, "Economic consequences of wildfire adaptation: Public safety power shutoffs in California," Available at SSRN 4417392, 2023.

14 M. Kreidler, "Liability insurance market conditions for Washington state electric utility companies," Jan. 1, 2023. Accessed: Dec 2023. Available: https://www.insurance.wa.gov/sites/default/files/documents/2023-utility-liability-study.pdf [Accessed: December 2023]

15 J. W. Mitchell, "Analysis of utility wildfire risk assessments and mitigations in California," *Fire Saf. J.*, vol. 140, p. 103879, 2023.

16 "ON FIRE: The report of the wildland fire mitigation and management commission," Sept. 2023. Accessed: Dec 2023. Available: https://www.usda.gov/sites/default/files/documents/wfmmc-final-report-09-2023.pdf [Accessed: December 2023]

17 A. Arab, A. Khodaei, R. Eskandarpour, M. P. Thompson, and Y. Wei, "Three lines of defense for wildfire risk management in electric power grids: A review," *IEEE Access*, vol. 9, pp. 61577–61593, 2021.

18 M. Finney and R. Koury, "Allstate also halting home insurance policies in California due to wildfire risk," *ABC7NEWS*. Available: https://abc7news.com/farmers-insurance-state-farm-allstate-home-wildfire/13335307/

19 NOAA National Centers for Environmental Information (NCEI) U.S. Billion-Dollar Weather and Climate Disasters, 2024. Accessed: Sep 2024. https://www.ncei.noaa.gov/access/billions/, doi: https://doi.org/10.25921/stkw-7w73.

20 M. Edesess, "Catastrophe bonds: An important new financial instrument," EDHEC-Risk Institute, Nice, France, 2014.

21 M. Galeotti, M. Gürtler, and C. Winkelvos, "Accuracy of premium calculation models for CAT bonds—An empirical analysis," *J. Risk Insur.*, vol. 80, no. 2, pp. 401–421, 2013.

22 M. A. Finney, "FARSITE: Fire area simulator-model development and evaluation," U.S. Dept. Agriculture, Forest Service, Rocky Mountain Research Station, 1988.

23 National Solar Radiation Database. Accessed: Dec 2023. Available: https://nsrdb.nrel.gov/

24 B. Sohrabi, A. Arabnya, M. P. Thompson, and A. Khodaei, "A wildfire risk-rating methodology for power grid infrastructure," *IEEE Access* (under review), 2024.

3

The Economics of Resilience-Centered Asset Management

3.1 Principles of Resilience-Centered Asset Management

Asset management comprises a set of engineering and management practices applied to valuable assets of a system to deliver the required level of service to the customers. Asset management is a critical responsibility of electric utility companies to maintain power network reliability and quality of service at acceptable levels by reducing the failure probability of critical grid components. In other words, asset management extends the lifetime of equipment and decreases the risk of equipment failure and unplanned power outages. Effective asset management of electric power systems has a great impact on the quality of services and significantly contributes to reducing the total operation costs of the utilities in the face of climate shocks and natural disasters. The effects of adverse weather conditions on the reliability of power systems is an undeniable fact that has received increasing attention from utility companies and research communities in recent years.

There is a vast literature on asset management of power infrastructure. However, this chapter introduces several of the most relevant research work. In maintenance scheduling, [1] developed a stochastic model for the optimal long-term maintenance scheduling of transmission lines and generation units coordinated with short-term security-constrained unit commitment. Random contingencies of transmission lines and generation units, load forecast errors, and fuel price fluctuations were simulated as scenario trees using the Monte Carlo simulation. Lagrangian relaxation was used to decompose the problem into maintenance and stochastic unit commitment subproblems. The results demonstrated the potential savings in operation costs due to the proposed model. A two-stage maintenance optimization model in short-term and mid-term horizons for a transformer was presented in [2] and [3]. The proposed model is formulated as a mixed-integer

The Economics of Climate Resilience in Power Infrastructure, First Edition. Ali Arabnya and Amin Khodaei.

linear program. The actual and expected condition of the transformer, as well as the constraints on $N-1$ contingency of the system, are considered in the model. The results demonstrate the solution quality and computational efficiency of the proposed model. A comprehensive reliability-centered maintenance framework for power distribution systems was presented in [4] and [5]. The underlying concepts, algorithms, and mathematical models were described in [4], while the application of the proposed framework was illustrated in [5].

In the context of consideration of adverse weather effects in asset management of power infrastructure, [6] proposed a probabilistic method to assess the vulnerability of power distribution poles to hurricanes subject to potential effects of changing climate. Two methods for the design of distribution poles in the United States, i.e., the American Society of Civil Engineering's method and the method of the National Electric Safety Code, were used to analyze the vulnerability of a distribution pole subjected to hurricanes. The obtained risk index is used to improve the design and maintenance decisions for power distribution poles to ensure the adequacy of their strength to withstand wind gusts. In [7], a risk assessment method for infrastructure technology planning to improve the power supply resiliency to natural disasters was proposed. Reduced cost and power supply availability are considered two fundamental decision factors in this hurricane planning approach. Different aspects of adverse weather on reliability and asset management of power infrastructure are investigated in [8–11].

In the context of topological aspects of the power grid and its effects on hardening the system against hurricanes, [12] proposed two models to solve power systems blackout problems using mixed-integer programming. The optimization problems that prevent large-scale blackouts in transmission grids subject to stochastic damage scenarios are considered. The first model decides on selecting the transmission lines to be expanded in capacity to guarantee that after damage to transmission lines in different scenarios, all power flows are within desired capacities. The second model considers the dynamics of cascades to find an optimal reinforcement plan that can passively survive a potential cascade. In [13], hurricane damage predictions and topological assessment to characterize the impact of hurricanes on the reliability of power systems were combined. Component fragility models are applied to simultaneously predict failure probability for individual transmission and distribution power network elements. The damage model is calibrated using power network component failure data for Hurricane Ike's strike at the Greater Houston area in 2008. This research demonstrates that storm hardening activities need to consider topological features, such as network meshedness, centrality, and clustering, and the compact irregular ring mesh topology.

In the context of the cost-effectiveness of power infrastructure hardening, [14] proposed a probabilistic model for analyzing electric power infrastructure risk mitigation investments. A parametric model is proposed to evaluate the trade-offs between wetland restoration and infrastructure hardening for the power grid. A hybrid economic input-output life-cycle analysis is used to analyze the model. The result indicates that wetland restoration and undergrounding of power infrastructure are not preferred over keeping them without wetland protection. Reference [15] proposed a system dynamics-based analysis of the cost-effectiveness of hurricane mitigation strategies of power distribution poles. Furthermore, it illustrates the systemic and dynamic nature of long-term maintenance and replacement strategies on the systems' cost-effectiveness and performance under hurricane effects.

In the context of emergency response planning for power systems, in [16], the research problems and models for substations and/or distribution feeders planning under normal and emergency conditions are reviewed and discussed. A case study on hurricane planning and rebuilding the electrical infrastructure along the Gulf Coast for Hurricane Katrina is presented in [17]. In [18], a stochastic integer program is proposed to find the optimal schedule for inspection, damage evaluation, and repair in the post-earthquake restoration of an electric power system. The aim is to minimize the time that each customer is without power on average. A comprehensive survey of models and algorithms for emergency response logistics in electric distribution systems, including reliability planning with fault considerations and contingency planning models, are presented in [19] and [20].

As a maintenance strategy, frequent inspection of civil infrastructure, including power systems, is neither practical nor cost-effective. Reference [21] proposes a Partially Observable Markov Decision Process (POMDP) model for inspecting, maintaining, and repairing civil infrastructures. The proposed model minimizes discounted life-cycle costs of the infrastructure. The decision variables determine when and how to inspect and repair the infrastructure. The formulated problem is solved by modifying the original one-pass algorithm, which provides exact solutions. In another work [22], the same authors present the theory and algorithm of the POMDP and its application in the maintenance management of civil infrastructure. An algorithm is established, and related topics in structural maintenance management are discussed.

Other research demonstrates the advantage of applying the POMDPs in the asset management of power systems. The most relevant research work includes [23] and [24]. Reference [23] studies the optimal preventive maintenance strategies for wind turbines by considering their operations under stochastic weather conditions. The problem is modeled as POMDP in an infinite horizon. The weather constraints, the lengthy lead time to deliver the maintenance services, and generation loss due to downtime are incorporated into the model. Furthermore, several structural properties, i.e., a set of closed-form expressions for the optimal policy are derived in their study. In [24], a wind turbine system with multi-level deterioration and multiple failure modes is studied. The season-dependent condition-based maintenance is modeled using POMDP with heterogeneous parameters. A backward dynamic programming algorithm is used to solve the model numerically. The results indicate that the dynamic condition-based maintenance strategy significantly improves costs and reliability compared to the static model and the current industry practices. Authors in [25] proposed a mixed-integer non-linear programming (MINLP) model that integrates a risk-based index to enhance resilience of distribution poles in the face of hurricanes. The proposed model takes into account the likelihood of pole failures and the resulting power outages. The results indicate that optimized maintenance planning can improve resilience throughout the system's life cycle. Reference [26] introduced a planning model for electrical transmission grids during hurricanes using meteorological data and fragility curves of grid components. The proposed model generates a set of scenarios for short- and long-term planning, including for emergency asset management. A case study of Hurricane Dolly affecting a synthetic 2000-bus test system in Southern Texas was presented to demonstrate its usefulness in resilience planning of power infrastructure in the face of hurricanes.

Analysis of the optimal policy's monotonicity and other structural properties can be utilized for heuristic use in the POMDP models and coping with the "curse of dimensionality" of this class of problems. As one of the most relevant research works, [27] studies the structured maintenance policies on interior sample paths. The problem of adaptively scheduling maintenance actions for a multi-state deteriorating system with hidden failure is modeled as POMDP. The structural properties of the optimal policy to minimize the expected maintenance cost are extended in their study. In another study [28], maintenance policies for systems with condition monitoring and obvious failures are investigated. The problem of adaptively scheduling perfect and imperfect observations and preventive maintenance actions for a system with Markovian deterioration and obvious failures is modeled. The POMDP is used to formulate the problem to minimize the expected long-run average cost per unit of time. The structural properties to devise a closed-form heuristic for the case with perfect information are derived and adjusted for the case with imperfect information.

Despite the given attention to weather effects, a holistic view of the global and long-term climatological effects of phenomena such as El Niño/La Niña is lacking in the power systems literature and perhaps in practice. According to the National Oceanic and Atmospheric Administration [29], "El Niño is characterized by unusually warm ocean temperatures in the Equatorial Pacific, as opposed to La Niña, which is characterized by unusually cold ocean temperatures in the Equatorial Pacific. El Niño is an oscillation of the ocean-atmosphere system in the tropical Pacific having important consequences for weather around the globe."

Several interesting works in actuary science and risk theory research communities have considered this issue in hurricane arrival analysis. The most relevant research work is [30], which considers the seasonal effects and El Niño/La Niña phenomenon to model the hurricane arrival times. The non-homogeneous Poisson processes are used for modeling the fluctuations of hurricane arrivals due to seasonality in intensity function due to the El Niño/La Niña phenomenon.

In our previous work [31] and [32], reactive and proactive restoration strategies for power system infrastructure considering hurricane effects were proposed, respectively. In [33], we proposed a dynamic maintenance scheduling for power systems considering hurricane effects. Our initial motive for this chapter is from the study by Edison Electric Institute [34] indicating that "infrastructure hardening should not come only as a result of storm damage and teardowns but as part of a regular maintenance schedule." To this end, this chapter investigates the long-term dynamics of the system by integrating the restoration and hardening strategies with commonly used maintenance strategies, i.e., inspection, preventive maintenance, and corrective maintenance. We use POMDP to develop a generic model for integrated condition-based maintenance and infrastructure hardening of the major components of the electric power systems infrastructure. In our modeling paradigm, two different hardening strategies, namely with temporary and permanent effects (during the life cycle of the infrastructure), are modeled as a generalized model. The hardening strategy with permanent effects is a special case of the proposed generalized model. Independent homogeneous Poisson processes are used to model the long-term climatological effects of El Niño/La Niña as a piecewise function to be incorporated in the hurricane survival function. The derived function is used to compute the transition probability of the failure due to a hurricane. While the transition probabilities to the damaged state due to hurricanes are non-stationary, the deterioration of the system and the subsequent failure are considered stationary. The proposed model in this chapter intends to provide informed decisions on inspection schedules, preventive maintenance actions with respect to the condition of the system, and a cost-effective infrastructure hardening schedule in a finite planning horizon. The model is illustrated with a hurricane as a climate shock to a critical component of the power systems, i.e., a standard oil-filled transformer.

3.2 Economic Variables for Resilience-Centered Asset Management

Consider a critical component of power system infrastructure located in a hurricane-prone area. From this point forward, this infrastructure component is alternatively called the *system*. The system is subject to breakdown due to two independent and competing random failure processes: failure due to maximum deterioration and failure due to hurricane strikes. In addition to the failure risks, the system is subject to progressive deterioration over time. As the system deteriorates over time, the probability of failure increases. Furthermore, the associated cost of bringing the system to *as-good-as-new* condition increases. On the other hand, regardless of the deterioration condition of the system, when a hurricane strikes, it can result in system failure. However, the risk of a hurricane strike varies from one season to another and from one year to another year due to the periodic nature of the phenomenon and the long-term climatological effects of El Niño/La Niña on hurricane arrivals [30]. Therefore, the overall risk of system failure over the planning horizon changes dynamically. In addition, the downtime cost of the system can be non-stationary due to variations in demand for electricity and generation cost over the planning horizon. The trade-off arises between the risk of failure, the associated expected downtime cost of the system, and the capital invested in asset management.

Using online data from the Supervisory Control and Data Acquisition (SCADA) system, the failure condition of the power system infrastructure can be detected. It is assumed that the failure due to hurricane strikes can be fully observed and distinguished from failure due to maximum deterioration. On the other hand, the deterioration level of the system can be *partially observed* from the probabilistic reliability function or condition monitoring systems, given the system is still in a functional state. To fully observe the deterioration level of the system, inspections are required. However, frequent inspections of the system are neither practical nor cost-effective. Therefore, a combination of full and partial observations can be utilized to make maintenance and system-hardening decisions.

3.2.1 State Space

We consider a two-dimensional state (s, z) to describe the state of the system. The first element (s) represents the deterioration level of the system, while the second element (z) shows the hardening state of the system. When $s \in \{1, \dots, k\}$, the system is functional. As s increases from 1 to k, the system deteriorates. That is, $s = 1$ indicates that the system is in the *as-good-as-new* condition, while $s = k$ indicates the most deteriorated functional state of the system. We also consider two *failure* modes (states) for the system: failure due to maximum deterioration, denoted by $s = k+1$, and failure due to damage by the hurricane, denoted as $s = k+2$. Figure 3.1 shows the transition diagram for the first element of the state space.

The second element of the system's state, i.e., z, is a non-negative integer, which represents the number of time periods that the system will be in the *hardened* state. Variable z can be interpreted as a countdown timer for hardening measures that are of a temporary nature. For instance, tree trimming as a hardening measure will

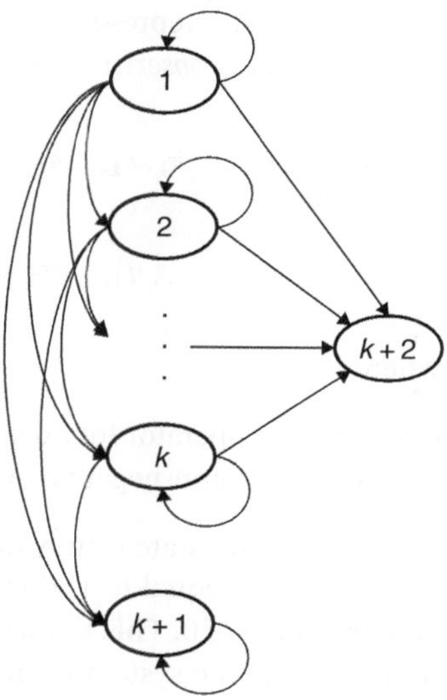

Figure 3.1 Transition diagram for the first element of the state space.

be in effect for a few years and must be repeated after a certain period, depending on geographical and other environmental factors. When the hardening measure is taken, the hardening state for the system is immediately brought to its maximum value n, i.e., the average length of time during which the hardening measure will be in effect. By passing each time period after the last hardening, the value of z is reduced by one unit. When the effect of the last hardening measure expires (which takes n time periods from the last hardening), then we have $z = 0$. Regardless of the system's deterioration level and the hardening measures taken, the system can fail due to a hurricane if it cannot withstand the hurricane's wind forces.

Overall, the *original state space* of the system is defined as follows:

$$S = \{(s, z);\quad s \in \{1, \ldots, k+2\};\quad z \in \{0, \ldots, n\}\} \tag{3.1}$$

Without a full observation, the exact deterioration state of a functional system cannot be known. On the other hand, frequent system inspection is not practical in many cases involving power infrastructure. One approach to overcome this barrier is to incorporate the probability of being in a particular state at each decision epoch. To model this uncertainty, a probability distribution can be used to define the system's state as the decision maker's belief about the underlying state of the system. To this end, the state of the systems is defined as follows:

$$(\boldsymbol{\pi}, z) = ([\pi_1, \ldots, \pi_k, \pi_{k+1}, \pi_{k+2}], z) \tag{3.2}$$

where π_i is defined as the probability that the system is in state i with $i \in \{1, \ldots, k+2\}$, while $\boldsymbol{\pi}$ is known as *information state* in the literature [27]. The combination of the first element (information state) and the second element (hardening state) represents the system's state. The state space under the POMDP, i.e., the *partially observable state space*, is defined as follows:

$$S' = \left\{ ([\pi_1, \dots, \pi_k, \pi_{k+1}, \pi_{k+2}], z); 0 \leq \pi_i \leq 1, \quad \forall i; \quad \sum_{i=1}^{k+1} \pi_i = 1 \oplus \pi_{k+2} = 1; \right.$$
$$\left. z \in \{0, \dots, n\}, \quad n \in \mathbb{N}_0; \right\} \tag{3.3}$$

where

$\oplus$ is the logic operator for *exclusive or*
$\mathbb{N}_0$ is the set of non-negative integers.

If the system's state is fully observable and known, i.e., an element of the information state is equal to 1, then the remaining elements of the information state take the value of 0. This known state is called an *extreme state*, e_i. For instance, e_1 indicates that the system is known to be in the *as-good-as-new* condition or e_{k+1} indicates that the system is certainly in a state of failure due to deterioration.

When the system is functional, then $\sum_{i=1}^{k} \pi_i = 1$. When the system is non-functional, it is either in failure state due to deterioration ($\boldsymbol{\pi} = e_{k+1}$), or in the failure state due to hurricanes ($\boldsymbol{\pi} = e_{k+2}$). Since both the failure states are fully observable, there are no probabilities associated with them in the partially observable state space when the system is functioning. Our policy is to immediately perform the *corrective maintenance* when the system fails due to deterioration and immediately perform *restoration* when it fails due to hurricanes. Therefore, when the system is functional, we always have $\pi_{k+1} = 0$ and $\pi_{k+2} = 0$.

3.2.2 Hurricane Survival Modeling

The strength of power structures and the stress of wind gusts in hurricane strikes can be modeled by the lognormal distribution [13, 35]. Assuming that μ is the mean of a normal variable and σ^2 is the variance of a normal variable, Consider that the wind gust speed is a random variable distributed as $G \sim LogNor\left(\mu_G, \sigma_G^2\right)$. The maximum wind gust speed that the system can withstand is a random variable distributed as $G' \sim LogNor\left(\mu_{G'}, \sigma_{G'}^2\right)$ when the system is not in the hardened state ($z = 0$); and $G'' \sim LogNor\left(\mu_{G''}, \sigma_{G''}^2\right)$ when the system is in hardened state ($z = 0$). The following definitions and assumptions are needed to continue the discussion.

Definition 3.1 Suppose X and Y are normally distributed. If X has a smaller variance or greater mean than Y, then random variable X *second order stochastically dominates (SSD)* random variable Y, denoted as $X \succeq_{SSD} Y$[36].

Assumption 3.1 When the system is in a hardened state, then the mean of its strength against wind stress is higher than when it is in a non-hardened state $(\mu_{G''} > \mu_{G'})$

Assumption 3.2 After hardening the system, the variance of strength does not increase $(\sigma^2_{G''} \leq \sigma^2_{G'})$.

Proposition 3.1 Random variable G'' second order stochastically dominates (SSD) random variable G', denoted as $G'' \succeq_{SSD} G'$.

Proof: As both G' and G'' are lognormally distributed, considering Assumptions 1 and 2, the relationship between corresponding means and between corresponding variances of normally distributed random variables $\ln(G)$ and $\ln(G'')$ still holds. Since $\ln(G'') \succeq_{SSD} \ln(G'')$, then $G'' \succeq_{SSD} G'$.

Assumption 3.3 The arrival of any two hurricane strikes and their associated wind gust speeds are considered to be independent of each other.

Using a dynamic stress-strength model in which both stress (i.e., the wind gust speed) and strength (i.e., the maximum wind gust speed that the system can withstand) are random variables, the probability that the system is not affected by hurricane strikes during period t, i.e., from decision epoch t until decision epoch $t+1$ is modeled as follows:

$$H_t^0 = \sum_{m=0}^{\infty} P\left\{G_1 < G_1', G_2 < G_2', \ldots, G_{N_t(\tau)} < G'_{N_t(\tau)}, \mid N_t(\tau) = m\right\} \ldots$$

$$P(N_t(\tau) = m) = \sum_{m=0}^{\infty} [P(G < G')]^m P(N_t(\tau) = m) \tag{3.4}$$

where

H_t^0 is the probability of survival in a non-hardened state
$N_t(\tau)$ is the number of hurricane strikes in period t

The annual seasonal pattern and long-term periodic behavior of hurricane arrivals considering El Niño/La Niña effects must be addressed using an appropriate model. Therefore, the probability of hurricane arrivals from the beginning of the planning horizon until time $t' = \tau t$ can be modeled using a non-homogeneous Poisson (NHP) process as follows:

$$P(N(t') = m') = \frac{\exp(-\Lambda(t'))(\Lambda(t'))^{m'}}{m'!} \tag{3.5}$$

where $\Lambda(t')$ is the cumulative intensity function for the NHP process.

Reference [30] derived $\Lambda(t')$ for a double-beta periodic intensity model which describes the effects of El Niño/La Niña on hurricane arrivals. By replacing $N_t(\tau)$ with $N(t')$ in (3.4), the reliability function of the system against hurricanes considering El Niño/La Niña is obtained. However, this approach is useful for risk analysis of a power system component when memoryless property is not desired. Another alternative approach is to directly use the average hurricane arrival rate in each period t within the long-term El Niño/La Niña cycle $c \gg t$, *i. e.*, λ_t and use a homogenous Poisson process to model the hurricane arrival rate during each long-term El Niño/La Niña cycle in a piecewise manner. In the latter approach, the survival probability of the system against a hurricane is independent of previous periods; hence, the memoryless property for transition probabilities to failure state due to hurricane holds.

We can write:

$$P(G < G') = P\left(\frac{G}{G'} < 1\right) = P(\ln(G) - \ln(G') < 0) \tag{3.6}$$

where $\ln(G) \sim N\left(\mu_G, \sigma_G^2\right)$ and $\ln(G') \sim N\left(\mu_{G'}, \sigma_{G'}^2\right)$. By standardizing these random variables and plugging the Poisson probability mass function in the hurricane survival function (3.4), we have

$$H_t^0 = \sum_{m=0}^{\infty}\left[\Phi\left(\frac{\mu_{G'} - \mu_G}{\sqrt{\sigma_G^2 + \sigma_{G'}^2}}\right)\right]^m \frac{\exp(-\tau\lambda_t)(\lambda_t\tau)^m}{m!} \tag{3.7}$$

where $\Phi(\cdot)$ is the cumulative distribution function of the normal distribution. Using the following geometric series

$$\sum_{q=0}^{\infty}\frac{r^q}{q!} = \exp(r) \tag{3.8}$$

we obtain

$$H_t^0 = \exp\left[\tau\lambda_t\left(\Phi\left(\frac{\mu_{G'} - \mu_G}{\sqrt{\sigma_G^2 + \sigma_{G'}^2}}\right) - 1\right)\right] \tag{3.9}$$

This function computes the probability of not being affected by hurricanes in period t, given the system is not in a hardened state ($z = 0$). In the same manner, the probability of survival of the system in period t when it is in hardened state ($z > 0$), denoted as H_t^1 is modeled by replacing $\mu_{G'}$ and $\sigma_{G'}^2$ with $\mu_{G''}$ and $\sigma_{G''}^2$, respectively. Therefore, the transition probability of failure due to hurricanes in period t, depending on the hardening state of the system which can be either $z = 0$ or $z > 0$, is obtained as $1 - H_t^0$ or $1 - H_t^1$, respectively.

3.2.3 Transition Probabilities

First, suppose that the system is currently functional with an information state $\boldsymbol{\pi}$. The probability that the system will survive until the next decision epoch, given there is no hurricane effect, is referred to as the *conditional reliability* function $R(\boldsymbol{\pi})$. This function is defined as follows:

$$R(\boldsymbol{\pi}) = \sum_{i=1}^{k} \sum_{j=1}^{k} \pi_i \, p_{ij} \tag{3.10}$$

where

p_{ij} is the transition probability from state i to state j in the original state space S
$\sum_{j=1}^{k+1} p_{ij} = 1, \quad \forall i \in \{1, \dots, k+1\}$

As shown by [28], based on the law of conditional probability, the information state at the beginning of the next decision epoch, given the system will still be functional, is

$$\tilde{\pi}_j(\boldsymbol{\pi}) = \begin{cases} \dfrac{\sum_{i=1}^{k} \pi_i p_{ij}}{R(\boldsymbol{\pi})}, & j = 1, \dots, k, \\ 0, & otherwise. \end{cases} \tag{3.11}$$

Therefore, the system makes a transition to the next information state with probability $R(\boldsymbol{\pi})$ as follows:

$$\tilde{\pi}(\boldsymbol{\pi}) = \{[\tilde{\pi}_1(\boldsymbol{\pi}), \tilde{\pi}_2(\boldsymbol{\pi}), \dots, \tilde{\pi}_k(\boldsymbol{\pi}), 0, 0]; \quad 0 \le \tilde{\pi}_i(\boldsymbol{\pi}) \le 1, \ \ \forall i = 1, \dots, k\} \tag{3.12}$$

If the current information state is $e_{k+2}\boldsymbol{\pi}$, the probability that the system fails due to deterioration (if there is no hurricane effect) is obtained as follows:

$$F(\boldsymbol{\pi}) = \sum_{i=1}^{k} \pi_i p_{i,k+1} = 1 - R(\boldsymbol{\pi}) \tag{3.13}$$

Therefore, given the system is not affected by hurricanes, with probability $F(\boldsymbol{\pi})$ it makes a transition to the *extreme* state e_{k+1}.

Regardless of the deterioration state of the system, it can be damaged by hurricane strikes. As derived in the previous subsection, in each period t, depending on the hardening state of the system, i.e., either $z = 0$ or $z > 0$, the functional state of the system makes a transition to the *extreme* state e_{k+2} with probabilities $1 - H_t^0$ or $1 - H_t^1$, respectively.

Recall that the second element of the system's state (the hardening state) is used as a counter, which shows the number of remaining periods in which the hardening action will be in effect. Therefore, if $z = u$ at decision epoch t, then regardless of the information state, with probability one, z makes a transition to $z' = max\,(0, u-1)$ denoted by $(u-1)^+$ at decision epoch $t+1$.

3.2.4 Action Space

The following actions constitute the action space at each decision epoch:

- No action (NA);
- Inspection (IN);
- Preventive maintenance (PM);
- Corrective maintenance (CM);
- Restoration (RS); and
- Hardening (HH).

When the system is functional, two actions are always available at each decision epoch: *no action* (*NA*) and *inspection* (*IN*). When the inspection action is taken, the exact functional state of the system is revealed. Subsequently, following each inspection, the decision maker can choose to either take *no action* or perform *preventive maintenance* (PM_i) based on the condition of the system (i.e., the revealed deterioration level i), denoted by the extreme information state e_i. By performing preventive maintenance, the system's reliability is fully restored, and the system is returned to the *as-good-as-new* condition. The immediate cost of preventive maintenance increases in the deterioration state of the system, i.e., $C_{PM_k} > C_{PM_{k-1}} > \cdots > C_{PM_1}$. Obviously, zero cost is associated with *NA* and C_{PM_1} actions. Therefore, a sub-action space, when it is revealed by *inspection* that the system is in deterioration state i, is represented by $\{NA, PM_i\}$ where $i \in \{1, \ldots, k\}$.

Due to the high cost of the outage and its consequences on system-level reliability of the grid, once the system goes to failure state e_{k+1}, we immediately perform *corrective maintenance*, denoted by *CM*, at an immediate cost of C_{CM}. Likewise, when the system fails due to hurricane damage, we instantaneously restore the system by taking *restoration* action, denoted by *RS*. The cost associated with the restoration is C_{RS}, where $C_{RS} \geq C_{CM} \geq C_{PM_k}$. Furthermore, to quantify the changing downtime cost, which is associated with the breakdown of the system during different periods of the year, a separate downtime cost, which is a function of time, can be considered for the system.

Finally, in each decision epoch, there is another type of action available, i.e., *hardening* action (*HH*), at the cost of C_{HH} strengthening the system against upcoming hurricanes. This action does not fully eliminate the damage risk due to hurricane strikes but significantly reduces the risk of such damages. The hardening strategy and its subsequent effects vary depending on infrastructure and geographical location. These strategies include but are not limited to vegetation management, undergrounding of distribution and transmission lines, distributed generation, and modernization of the smart grid [37]. Two types of effects, i.e., *temporary* and *permanent*, can be perceived for the hardening strategy. For instance, vegetation trimming, considering the growing nature of the trees from one side, and environmental protection and regulatory issues from the other side will be in effect for a certain period of time. Therefore, such a hardening action with *temporary* effects must be scheduled and repeated at specific intervals [37]. On the other hand, engineering solutions such as undergrounding of transmission lines or installing spacer cables and Cutout-Mounted Reclosers [38] result in *permanent* effects during the life-cycle of the infrastructure.

3.3 Economic Asset Management Model

3.3.1 Problem Formulation

The problem is formulated as a finite horizon POMDP. To construct the optimality equation, each action's expected cost must be specified first. The optimality equation for the proposed model is constructed as follows:

$$V_t(\pi, z) = \begin{cases} CM_t(e_{k+1}, z), & \pi = e_{k+1}, \\ RS_t(e_{k+2}, z), & \pi = e_{k+2}, \\ min\{NA_t(\pi, z), IN_t(\pi, z), HH_t(\pi, z)\}, & otherwise. \end{cases} \tag{3.14}$$

where $V_t(\pi, z)$ is the minimum total expected cost-to-go when the system is in information state π and hardening state z at decision epoch t.

As shown, when the system is in the fully observable information states, i.e., the *failure* due to hurricane (e_{k+1}), or *failure* due to aging and degradation (e_{k+2}), the corresponding actions are immediately taken to bring the system back to the *as-good-as-new* condition. When the system is functional, the most cost-effective action among NA, IN, and HH needs to be taken in that decision epoch.

When *no action* is taken at decision epoch t, given the state of the system is (π, z), the expected cost-to-go can be written as:

$$NA_t(\pi, z) = W_t(\pi, z) \tag{3.15}$$

where

$$\begin{aligned} W_t(\pi, z) = {} & \left[H_t^1 I(z > 0) + H_t^0 I(z = 0)\right] \\ & \times \beta[R(\pi)V_{t+1}(\tilde{\pi}(\pi), (z-1)^+) + F(\pi)V_{t+1}(e_{k+1}, (z-1)^+)] \\ & + \left[1 - \left(H_t^1 I(z > 0) + H_t^0 I(z = 0)\right)\right] \beta V_{t+1}(e_{k+2}, (z-1)^+) \end{aligned} \tag{3.16}$$

where

$(z-1)^+$ is the positive part of $(z-1)$ defined as $f(x)^+ = max(f(x), 0)$
$I(\cdot)$ is the indicator function

The first term in (3.16) represents the total expected cost in period t if the system is not affected by hurricanes. It considers two possible outcomes, which can result either in continuing the system's functionality until period $t + 1$, or system failure due to deterioration. The second term in (3.16) presents the expected cost in period t if the system fails due to a hurricane. We assume that the failure due to a hurricane overrides any other states of the system and takes the information state into e_{k+2}.

If the *hardening* measures are taken, they do not guarantee that the system will survive against hurricanes but significantly reduce the chance of failure if the hurricane strikes. The expected cost-to-go by taking the *hardening* action is as follows:

$$HH_t(\boldsymbol{\pi}, z) = C_{HH} + H_t^1 \beta \left[R(\boldsymbol{\pi})V_{t+1}(\tilde{\boldsymbol{\pi}}(\boldsymbol{\pi}), n-1) + F(\boldsymbol{\pi})V_{t+1}(e_{k+1}, n-1)\right] + \left(1 - H_t^1\right) \beta V_{t+1}(e_{k+2}, n-1) \tag{3.17}$$

where n is the number of periods that hardening will be in effect. By taking the *hardening* action, an immediate cost of C_{HH} is incurred, and the system continues to evolve. Since hardening is performed discretely from the system, we do not consider a downtime cost associated with this action. As shown in the second and third terms of (3.17), by hardening the system, the hurricane survival probability in period t immediately increases to H^1, so the risk of being affected by a hurricane is reduced. By taking this action in decision epoch t, the hardening state of the system instantaneously makes a transition to n with probability 1.

Once the system fails due to the hurricane, the *restoration* action must be taken immediately. Assuming DT_t^a is downtime cost in period t due to action a, the expected cost-to-go associated with the *restoration* action is given by:

$$RS_t(e_{k+2}, z) = C_{RS} + DT_t^{RS} + W_t(e_1, z) \tag{3.18}$$

By restoration, the system is brought to the *as-good-as-new* condition. As shown in (3.18) by taking the *restoration* action, the restoration cost as well as downtime cost are immediately incurred to the system. In addition, $W_t(e_1, z)$ which represents the expected cost-to-go based on the evolution of a renewed system at decision epoch t is incurred. In our model, without loss of generality, we assume that renewing the system by taking the *restoration* action does not impact the system's hardening state.

When the system fails due to deterioration, it needs to be repaired immediately. The *corrective maintenance* action is taken to bring the system to the *as-good-as-new* condition. The expected cost-to-go when the *corrective maintenance* action is taken is as follows:

$$CM_t(e_{k+1}, z) = C_{CM} + DT_t^{CM} + W_t(e_1, z) \tag{3.19}$$

where the first term is the corrective maintenance cost, and the second term is the downtime cost. Like (3.18), the third term in (3.19) represents the expected cost-to-go from period t for a system in an *as-good-as-new* condition.

With transition probabilities, the deterioration level of a functional system can be partially observable. The only way to reveal the exact deterioration state of a functional system is to perform the *inspection* action. However, inspection of the system at each decision epoch is not cost-effective. Therefore, a sound decision about the inspection of the system is required to be made. After inspection, with respect to the condition of the system, a proper decision regarding preventive maintenance action can be made. Therefore, the expected cost-to-go associated with taking the inspection action is obtained as follows:

$$IN_t(\boldsymbol{\pi}, z) = C_{IN} + \sum_{i=1}^{k} ACT_t(e_i, z)\pi_i \tag{3.20}$$

where

$$ACT_t(e_i, z) = min\{NA_t(e_i, z), PM_t(e_i, z)\} \tag{3.21}$$

where

C_{IN} is the immediate inspection cost

$PM_i(e_i, z)$ is the expected cost-to-go of the corresponding *preventive maintenance* action.

As a result of preventive maintenance, the system is returned to the *as-good-as-new* condition. The expected cost-to-go is as follows (same as (3.19) except the immediate costs of maintenance and downtime):

$$PM_t(e_i, z) = C_{PM_1} + DT_t^{PM_i} + W_t(e_1, z) \tag{3.22}$$

The proposed formulation is a generic model for systems in which the hardening action has *temporary* effects. A special case of this generic formulation can describe the systems in which the hardening action has *permanent* effects during the system's lifecycle. To do so, simply the value for n needs to be chosen sufficiently large, i.e., $n \geq T$.

3.3.2 Solution Method

The optimality equations are solved using the backward induction algorithm [39]. The algorithm starts from the last period, and the optimal policy for all periods is sequentially constructed. To use backward induction to solve for optimal policy in the proposed POMDP model, we approximate the sample path denoted by Ω as *a priori*. The sample path is defined as a sequence of information states evolving by taking *no action* [28]. To this end, we need to construct a sample path from the initial belief state $\boldsymbol{\pi}^0$. Furthermore, the sample path that starts from each functional extreme information state needs to be constructed. Therefore, using (3.11), we construct $k+1$ sample paths denoted by $\Omega_{\pi^0}, \Omega_{e_1}, \ldots, \Omega_{e_k}$. As explained earlier, due to the high cost of an outage, our policy is to immediately perform *corrective maintenance* once the system goes to a *failure* state due to deterioration and *restore* once it goes to a *failure* state due to hurricanes. Therefore, once the system goes into the extreme states e_{k+1} or e_{k+2}, it will not evolve by taking any action. However, based on our policy, the system is immediately brought back to the *as-good-as-new* condition in such circumstances, so the system evolves on the sample path Ω_{e_1}. Hence, there is no need to construct the sample paths $\Omega_{e_{k+1}}$ and $\Omega_{e_{k+2}}$.

To construct a sample path from a starter information state, e.g., $\boldsymbol{\pi}$, using (3.11) we obtain the first element (i.e., an information state) of the sample path by $\tilde{\boldsymbol{\pi}}^1 = \tilde{\boldsymbol{\pi}}(\boldsymbol{\pi})$, and the second element by $\tilde{\boldsymbol{\pi}}^2 = \tilde{\boldsymbol{\pi}}(\tilde{\boldsymbol{\pi}}(\boldsymbol{\pi}))$, and so forth. A sample path Ω_{π} emanating from $\boldsymbol{\pi}$ is defined as $\{\tilde{\boldsymbol{\pi}}^1, \tilde{\boldsymbol{\pi}}^2, \dots, \tilde{\boldsymbol{\pi}}^L\}$, where $L = \min\{l : \|\tilde{\boldsymbol{\pi}}^l(\boldsymbol{\pi}) - \tilde{\boldsymbol{\pi}}^{l-1}(\boldsymbol{\pi})\| \leq \varepsilon\}$. The last information state in the sample path, i.e., $\tilde{\boldsymbol{\pi}}^L$ is an *absorbing state* in which from this point forward the deterioration of the system shows a stationary behavior [24], [28]. As it is proven in [40], as long as the discrete-time Markov chain is acyclic, there exists L for any $\varepsilon > 0$. By constructing the described sample paths for $\Pi = \{\boldsymbol{\pi}^0, e_1, \dots, e_k\}$, the set $\Psi(\Pi) = \bigcup_{\pi \in \Pi} \Omega_{\pi}$ includes all of the information states that the system can reach in its evolution. Therefore, without compromising the accuracy, the solution space is significantly reduced which results in higher computational efficiency of the POMDP solution algorithm.

Algorithm 3.1 shows the steps of the POMDP solution algorithm. As shown, the initial belief state and the extreme states play a central role in constructing the reduced information state space of the system. The cardinality of set Π is denoted

---

Algorithm 3.1 The POMDP Solution Algorithm

Step 1. Select an unchosen $\boldsymbol{\pi}' \in \Pi = \left\{\boldsymbol{\pi}^0, \mathbf{e}_1, \dots, \mathbf{e}_k\right\}$; Set $|\Pi| = |\Pi| - 1$.
Step 2. Set $\boldsymbol{\pi} = \boldsymbol{\pi}'$; and $l = 1$.
Step 3. Compute $R(\boldsymbol{\pi})$ and $F(\boldsymbol{\pi})$ using (3.10) and (3.13).
Step 4. Generate the new information state $\tilde{\boldsymbol{\pi}}^l(\boldsymbol{\pi})$ using (3.11).
Step 5. Set $\boldsymbol{\pi} = \tilde{\boldsymbol{\pi}}^l(\boldsymbol{\pi})$; and $l = l + 1$.
Step 6. If $\left\|\tilde{\boldsymbol{\pi}}^l(\boldsymbol{\pi}) - \tilde{\boldsymbol{\pi}}^{l-1}(\boldsymbol{\pi})\right\| \leq \varepsilon$ stop; Otherwise, return to step 3.
Step 7. Generate sample path Ω_{π}.
Step 8. If $|\Pi| > 0$, return to step 1.
Step 9. Set $\Psi(\Pi) = \bigcup_{\pi \in \Pi} \Omega_{\pi}$.
Step 10. Set $t = T$; Set $V_T(\boldsymbol{\pi}, z) = 0, \forall \boldsymbol{\pi} \in \Psi(\Pi), \forall z \in \{0, \dots, n\}$.
Step 11. Set $t = t - 1$; Compute H_t^0 and H_t^1 using (3.9); Using (3.14) compute $V_t(\boldsymbol{\pi}, z), \forall \boldsymbol{\pi} \in \Psi(\Pi), \forall z \in \{0, \dots, n\}$; Set optimal policy $A_t^*(\boldsymbol{\pi}, z) = \arg\min V_t(\boldsymbol{\pi}, z)$.
Step 12. If $t = 1$, stop; Otherwise, return to step 11.

---

by $|\Pi|$ in the algorithm. Steps 1–8 construct the sample paths emanating from each information state member of the set Π. These steps are iterated until the corresponding sample paths emanating from every member of the set Π are obtained. Note that in step 6, the Euclidean norm of the difference of two consecutive information states emanating from information state π is computed and compared with the small value ε. When this norm becomes smaller than ε, the associated sample path Ω_π reaches its steady state behavior and no longer evolves. In step 9, the reduced information state space of the system, i.e., $\Psi(\Pi)$ is formed, where $\Psi(\Pi) \subset S'$. Step 10 incorporates the terminal values of the system at the end of the planning horizon at decision epoch T. We assume that the system will be overhauled at the end of the planning horizon regardless of its condition. Therefore, without loss of generality, the terminal values for all states are assumed as $r_T(\pi, z) = 0$.

Step 11 computes the minimum expected cost-to-go and finds the optimal policies in a backward manner. This step is iterated until the optimal actions for all decision epochs in the entire planning horizon are obtained. As explained, an MDP equivalent of the POMDP problem with reduced state space is solved to find the optimal solution for the proposed model. This approach enables one to solve a non-stationary problem for an exact solution. This conversion of the POMDP to its MDP equivalent results in the inflation of the problem size. However, this inflation has already been minimized through steps 1–9, as explained earlier. As shown, a considerable number of states and associated computations are eliminated without significant compromise on the accuracy of the model.

3.4 Case Study

In this section, we analyze a standard oil-filled transformer to illustrate the application and effectiveness of the proposed model. Dissolved Gas-in-Oil Analysis (DGA) is widely used in the power industry as a useful diagnostic tool for evaluating the health condition of transformers. Measuring certain gases generated in an oil-filled transformer is a reliable indicator of an existing malfunction that may result in failure if not corrected.

The failure mechanisms include severe overloading, pump motor failure, arcing, partial discharge, and low-energy sparking, among others. These failure mechanisms, which can occur singly or simultaneously, decompose the insulating materials and form a variety of combustible and non-combustible gases. Normal operation of a transformer can also result in the formation of various gases [41]. As shown in Table 3.1, based on key gas concentration limits, a four-level criterion was developed in [41] to identify the deterioration level of a transformer. Deterioration level 1 represents *as-good-as-new*, while deterioration

Table 3.1 Dissolved key gas concentration limits (μl/l(ppm)).

Key gas	Hydrogen (H_2)	Methane (CH_4)	Acetylene (C_2H_2)	Ethylene (C_2H_4)	Ethane (C_2H_4)	Carb. mono (CO)	Car. dio, (CO_2)	TDCG
Level 1	<100	<120	<35	<50	<65	<350	<2500	<720
Level 2	101–700	121–400	36–50	51–100	66–100	351–570	2500–4000	721–1920
Level 3	701–1800	401–1000	51–80	101–200	101–150	571–1400	4001–10,000	1921–4630
Level 4	>1800	>100	>80	>200	>150	>1400	>10000	>4630

Source: Ref. [41]/with permission of IEEE.

level 4 represents *failure* or near failure condition. Deterioration levels 2 and 3 represent the system with *minor* and *major* deteriorations, respectively. Note that TDCG in the last column stands for total dissolved combustible gas.

For our analysis, we use data from [42]. In their study, the same deterioration levels, and key gas concentration limits as [41] are considered, while condition 4 is considered a *failure* due to deterioration. By analyzing the historical data from condition monitoring of multiple transformers of the same type and manufacturer, they estimated the weekly transition probability matrix as follows:

$$P_w = \begin{bmatrix} 0.9917 & 0.0083 & 0.0000 & 0.0000 \\ 0 & 0.9936 & 0.0064 & 0.0000 \\ 0 & 0 & 0.9891 & 0.0109 \\ 0 & 0 & 0 & 1 \end{bmatrix} \tag{3.23}$$

Since the length of each period in our model is $\tau = 1$ month, we need to use a monthly transition probability in our analysis. Without loss of generality, we assume each month to be exactly four weeks. Therefore, the following 4-period (monthly) transition probability matrix for the underlying discrete-time Markov chain is used in our analysis:

$$P_m = \begin{bmatrix} 0.9672 & 0.0325 & 0.0003 & 0.0000 \\ 0 & 0.9746 & 0.0249 & 0.0005 \\ 0 & 0 & 0.9571 & 0.0429 \\ 0 & 0 & 0 & 1 \end{bmatrix} \tag{3.24}$$

where $P_m = P_w^4$.

To consider the long-term climatological effects of El Niño/La Niña, the problem needs to be solved for at least one long-term El Niño/La Niña cycle. From [30], we consider this cycle as $c = 5$ years. From the same study, we use the mean number of hurricane arrivals for each month t, denoted as λ_t during each El Niño/La Niña cycle for the Atlantic hurricanes, estimated from a data set of 102 years. On the other hand, the planning horizon needs to be long enough to study the economic impacts of capital expenditures on engineering solutions for the hardening strategy with *permanent* effects. Based on these considerations, we solve the problem for planning horizon of $T = 120$ months, i.e., for two consecutive El Niño/La Niña cycles. We also assume that the planning horizon starts at the beginning of the first El Niño/La Niña cycle.

The discount factor is assumed as $\beta = 0.99$, and the threshold value for the Euclidean norm as $\varepsilon = 0.01$. From [2], the inspection cost is considered as $C_{IN} = \$100$, the minor maintenance cost as $C_{PM_2} = \$1000$, the major maintenance cost as $C_{PM_3} = \$10,000$, and the corrective maintenance cost as $C_{CM} = \$100,000$. We set the restoration cost to $C_{RS} = \$100,000$ (same as the corrective maintenance cost). The minor maintenance is done online, while the major maintenance is performed offline [42]. Therefore, the downtime cost for minor maintenance is $DT_t^{PM_2} = 0$. In [42], the downtime costs for major and corrective maintenance have been aggregated into the corresponding maintenance cost. From [35], the mean of the logarithm of wind gust speed and its standard deviation are considered as $\mu_G = 3.85$ and $\sigma_G = 0.427$, respectively. The parameters associated with hurricane hardening actions vary based on geographical location and structural characteristics of the infrastructure in question. Without loss of generality, we assume the following parameters for our analysis: the cost of hurricane hardening action with *temporary* effects is assumed as \$3000, and the number of time periods that it will be in effect is assumed to be $n = 24$ months; the cost of hurricane hardening action with *permanent* effects is assumed to be \$20,000; the mean of the logarithm of structure's strength and its standard deviation *without* hardening are $\mu_{G'} = 4.7$ and $\sigma_{G'} = 0.1$, respectively; the mean of the logarithm of structure's strength and its standard deviation *with temporary* hardening effects are $\mu_{G''} = 4.9$ and $\sigma_{G''} = 0.1$, respectively; and, the mean of the logarithm of structure's strength and its standards deviation *with permanent* hardening effects are $\mu_{G''} = 5$ and $\sigma_{G''} = 0.1$, respectively. Table 3.2 summarizes the parameter values used for numerical analysis.

To illustrate the problem and the impacts of different asset management strategies, we analyze six cases as follows:

Case 1. Hurricane effects are considered. Preventive maintenance, corrective maintenance, and restoration are available for the system. In addition, tree trimming is used as a hardening strategy with *temporary* effects, i.e., $n = 24$ months.

Case 2. Hurricane effects are considered. Preventive maintenance, corrective maintenance, and restoration are available for the system. In addition, structural reinforcement is used as a hardening strategy with *permanent* effects.

Table 3.2 Parameter values for numerical analysis.

Parameter	Value	Source
β	0.99	Assumption
ε	0.01	Assumption
$C\text{-}_{IN}$	\$100	[40]
$C\text{-}_{PM2}$	\$1000	[40]
$C\text{-}_{PM3}$	\$10,000	[40]
$C\text{-}_{CM}$	\$100,000	[40]
$C\text{-}_{RS}$	\$100,000	Assumption
$C\text{-}_{HH}$ (temporary)	\$3000	Assumption
$C\text{-}_{HH}$ (permanent)	\$20,000	Assumption
DT_t^{PM2}	0	[40]
DT_t^{PM3}	Aggregated	[40]
DT_t^{CM}	Aggregated	[40]
DT_t^{RS}	Aggregated	Assumption
μ_G	3.85	[33]
σ_G	0.427	[33]
$\mu_{G'}$	4.70	Assumption
$\sigma_{G'}$	0.1	Assumption
$\mu_{G''}$ (temporary)	4.90	Assumption
$\sigma_{G''}$ (temporary)	0.1	Assumption
$\mu_{G''}$ (permanent)	5.00	Assumption
$\sigma_{G''}$ (permanent)	0.1	Assumption
λ_t (constant)	1.64	[28]
λ_t (variable)	Varying	[28]

Case 3. Hurricane effects are considered. Preventive maintenance, corrective maintenance, and restoration are available for the system. However, there is no hardening strategy.

Case 4. Hurricane effects are considered. Corrective maintenance and restoration are available for the system. However, preventive maintenance and hardening strategies are eliminated from asset management.

Case 5. There is no hurricane affecting the system. Preventive maintenance, corrective maintenance, and restoration are available for the system.

Case 6. There is no hurricane affecting the system. Only corrective maintenance is available for the system.

To analyze the implications of incorporating El Niño/La Niña, we analyze two scenarios for Cases 1–4, as follows:

Scenario I. The El Niño/La Niña effects on hurricane arrivals are considered. Therefore, non-homogeneous hurricane arrival rates λ_t for Poisson processes are used to describe the El Niño/La Niña effects on hurricane arrivals in the model.

Scenario II. The El Niño/La Niña effects are ignored. Instead, a homogeneous hurricane arrival rate for the Poisson process (i.e., $\lambda_t = 1.64$/month, $\forall t \in \{1, \dots, 120\}$) is used to model the hurricane arrivals.

We implemented Algorithm 1 in MATLAB and ran it for each case/scenario on a workstation with Core 2 Quad 2.33 GHz CPU and 8 GB memory. In addition to solving the problem, we simulate the obtained results from Scenario II of Cases 1–4 to evaluate the derived policies from a scenario with homogenous hurricane arrivals in an environment in which hurricanes follow a non-homogeneous arrival rate due to El Niño/La Niña phenomenon. The aim is to analyze the opportunity cost of ignoring the El Niño/La Niña effects to derive optimal policies.

In continuation, the numerical results on optimal policy for the system in the entire makespan of a long-term El Niño/La Niña cycle (i.e., $c = 5$ years) are described, and the dynamics of optimal policies in the hurricane and off-hurricane seasons during different years of the cycle are discussed.

In Scenario I of Case 1, for the first year of the cycle, it is optimal to take *no action* as long as the probability of being in state $1(\pi_1)$ is greater than 0.904 during hurricane season, while it is optimal to take *no action* when the probability of being in state 1 is greater than 0.875 for the rest of the first year. Once the probability of being in state 1 goes below the corresponding thresholds, it is optimal to *inspect* the system. Once inspection is performed, with respect to the revealed deterioration level of the transformer, the required maintenance needs to be delivered to bring the deterioration level of the transformer into state 1, i.e., *as-good-as-new* condition. The optimal period to perform infrastructure hardening (i.e., tree trimming) is in period 6, i.e., at the beginning of the first hurricane season in the El Niño/La Niña cycle. In the second year, it is optimal to perform an inspection once the probability of being in state 1 is less than 0.904, regardless of the time of the year. Since the hardening performed in the middle of year 1 is still in effect, no hardening action is required for year 2. In the third and fourth years of the cycle, as long as the π_1 is greater than 0.875, it is optimal to take no action; otherwise, the *inspection* and subsequent maintenance actions are required to be taken. The optimal time to perform the second infrastructure hardening action is at the beginning of the third year of the hurricane season. The third hurricane-hardening action is taken at the beginning of the hurricane season of the fifth year of the El Niño/La Niña cycle. The optimal *inspection/no action* decisions in the fifth year are similar to the first year of the cycle.

In Scenario II of Case 1, it is always optimal to perform an *inspection* once the probability of being in state 1 goes below 0.875; Otherwise, it is optimal to take *no action.* The first hardening action is performed at the beginning of the planning horizon. The second and third hardening actions are taken at the beginning of the third and fifth year, respectively.

In the first two years of El Niño/La Niña cycle in Scenario I of Case 2, it is optimal to do *inspection* once π_1 becomes less than 0.904 and 0.875 in hurricane and off-hurricane seasons, respectively. In the remaining three years of the cycle, it is always optimal to do an *inspection* once π_1 drops below 0.875. Otherwise, *no action* is the optimal strategy. The optimal time to perform the hardening action with *permanent* effect (i.e., structural reinforcement) is at the beginning of the hurricane season in the first year. In Scenario II of this case, it is always optimal to take *no action* as long as the probability of being in state 1 is greater than 0.875; otherwise, *inspection* is the optimal action. In this scenario, the hardening action is performed at the beginning of the planning horizon.

In the first, second, third, and fifth years of the cycle in Scenario I of Case 3, it is optimal to perform the *inspection* once π_1 is less than 0.875 and 0.904 in hurricane and off-hurricane seasons, respectively. However, in year 4, it is optimal to *inspect* once the probability of being in state 1 becomes less than 0.875. In Scenario II of Case 3, it is always optimal to do an inspection once π_1 becomes less than 0.904; Otherwise, it is optimal to take *no action.* As described, we have not taken any hardening measures in this case.

Table 3.3 shows the expected life-cycle cost of the transformer as well as the computation time in different cases and scenarios. As shown for Case 4, considering the El Niño/La Niña effects, the expected life-cycle cost of the transformer, when no preventive maintenance and infrastructure hardening measures are taken significantly increases by 175%, 301%, and 4.3% compared to Scenario I of Cases 1, 2, and 3, respectively. However, if the El Niño/La Niña effects are not considered (Scenario II), the reactive asset management strategy in Case 4 results in 183%, 312%, and 4.1% increase in expected life-cycle cost compared to Cases 1, 2, and 3, respectively.

Table 3.3 The expected life-cycle costs for cases 1–4.

Case #	Scenario 1	Scenario 2	Simulated model	Time (min)
1	$108,461	$109,023	$108,975	<10
2	$74,317	$74,785	$74,785	<250
3	$285,969	$296,406	$286,166	<3
4	$298,334	$308,580	$298,333	<3

Table 3.4 The expected life-cycle costs for cases 5 and 6.

Case #	Expected cost	Time (min)
5	\$5744	<2
6	\$37,830	<2

In Case 5, in which there are no hurricane effects (because of being located in a non-hurricane-prone area), the asset management strategy excludes hardening measures. The results indicate that similar to Scenario II of Cases 1, 2, and 3, in which the risk factor does not change over the planning horizon, the system needs to be *inspected* once the probability of being in an *as-good-as-new* condition (state 1) drops to below a threshold of 0.875. Otherwise, *no action* is required. In Case 6, for the same environmental circumstances described in Case 5, we analyze the system's life-cycle cost by not maintaining it unless it fails due to deterioration. As shown in Table 3.4, the system's life-cycle cost increases by 550% compared to Case 5.

Figure 3.2 illustrates the structure of optimal policy along the sample path for the system marked with probability thresholds of 0.875 and 0.904. As shown,

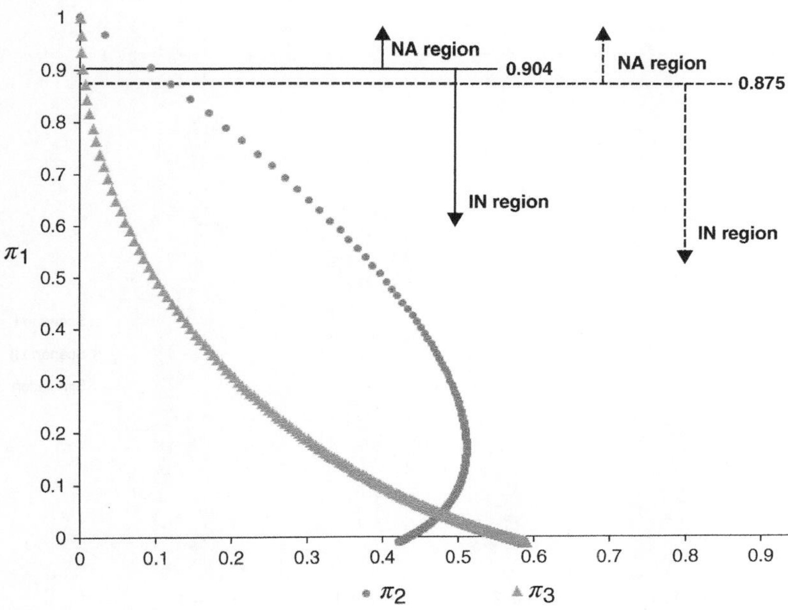

Figure 3.2 Structure of optimal policy on sample path.

probabilistically, it takes four months for a system in an *as-good-as-new* condition to reach the threshold of 0.904, while it takes five months to reach a probability threshold of 0.875. Therefore, if the transformer is not equipped with real-time condition monitoring devices, the corresponding time intervals for reaching these threshold values can be used to make decisions on maintenance planning for the transformer. For instance, in Case 6, it is optimal to inspect the transformer every five months, while in year 1 of Scenario I of Case 1, the optimal inspection time intervals from the last inspection are four and five months in hurricane and off-hurricane seasons, respectively.

We simulated the impacts of implementing the optimal policies from scenarios with homogeneous hurricane arrival patterns in an environment subject to El Niño/La Niña effects. The expected life-cycle cost of implementing such policies is $108,975, $74,785, $286,166, and $298,333 for Cases 1, 2, 3, and 4, respectively. Therefore, the expected loss/opportunity cost of implementing such inaccurate policies for Cases 1, 2, 3, and 4 are 0.005, 0.006, 0.036, and 0.034. As illustrated in Figure 3.3, the expected lifecycle of the transformer by implementing the derived policies from homogeneous hurricane arrival models in asset management strategies, which include full maintenance options and hardening measures, are not significant, while the strategies with no hardening measures are considerable.

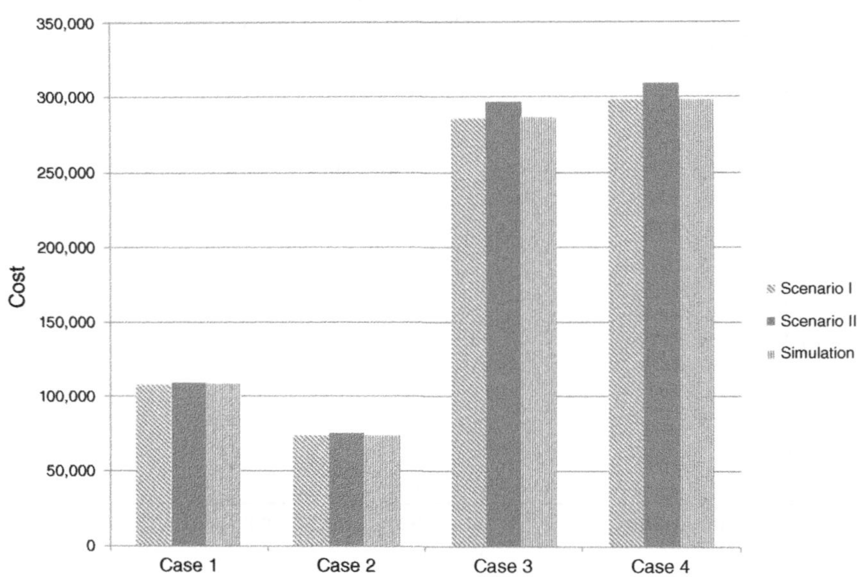

Figure 3.3 Comparison of expected cost-to-go and simulated model.

The cost of damage to the transformer due to the hurricane and the subsequent cost of interruption due to the long lead time of this critical asset is relatively high. Based on [42], from 1997 to 2001, the average property damage cost and the interruption cost of each damaged transformer in the United States were about \$1,736,586 and \$1,312,657, respectively. Even though in our study, we assumed a relatively low restoration cost for hurricane-related damages to the transformer (i.e., \$100,000), in all cases and scenarios, the optimal policy is to *harden* the system at every decision epoch in which there is a chance of a hurricane. This finding shows the importance of hardening measures in the cost dynamics of the asset management of critical power infrastructures in hurricane-prone areas.

Finally, as shown in Table 3.3, the computation time increases with the number of periods during which that hardening action will be in effect. In addition, as explained in Section III, the size of the information space has significantly been reduced in our model. For instance, the sample path emanating from extreme state e_1 reaches its steady state after 39 iterations (versus 120 iterations). Hence, the reduction in the size of the information space needs to be evaluated, as well as the improved computational efficiency of the proposed model.

3.5 Summary

In this chapter, we proposed a new model for asset management of power infrastructure to integrate long-term maintenance with hurricane hardening and restoration activities. We used the POMDP with a mixed-state space, i.e., a stochastic partially observable information state and a deterministic hardening state, as a framework to formulate the problem. Utilizing the long-run steady state behavior of system deterioration, we significantly reduced the size of information space, which resulted in improved computational efficiency of the backward induction algorithm without significantly compromising the accuracy of the optimal solution. The results indicate that the optimal maintenance decisions are sensitive to the hurricane seasons in El Niño/La Niña cycles. We simulated the derived optimal policies from systems with homogeneous hurricane arrival rates (which statistically are inaccurately distributed) in an environment with non-homogeneous hurricane arrival patterns (that are proven to be more accurately describing the reality), which considers El Niño/La Niña effects. The simulation results indicate that the opportunity cost for the system, when the derived policies based on homogeneous hurricane arrival assumptions are implemented, cannot always be significant. The reason can be the long-term and cyclic nature of the hardening strategies that remain in effect. However, using homogeneous hurricane arrival models to find the optimal policies can significantly overestimate the total life-cycle cost on the grid scale. This overestimation can lead to unrealistic budget constraints, prohibiting a cost-effective and productive asset management strategy at the utility company level. The results also demonstrate the importance of hardening strategy in asset management of the critical power infrastructure due to the considerable impacts of the failure due to hurricanes and the subsequent interruptions on the cost dynamics of the system.

References

1 L. Wu, M. Shahidehpour, and Y. Fu, "Security-constrained generation and transmission outage scheduling with uncertainties," *IEEE Trans. Power Syst.*, vol. 25, no. 3, pp. 1674–1685, Aug. 2010.

2 A. Abiri-Jahromi, M. Parvania, F. Bouffard, and M. Fotuhi-Firuzabad, "A two-stage framework for power transformer asset maintenance management—Part I: Models and formulations," *IEEE Trans. Power Syst.*, vol. 28, no. 2, pp. 1404–1414, May 2013.

3 A. Abiri-Jahromi, M. Parvania, F. Bouffard, and M. Fotuhi-Firuzabad, "A two-stage framework for power transformer asset maintenance management—Part II: Validation results," *IEEE Trans. Power Syst.*, vol. 28, no. 2, pp. 1395–1403, May 2013.

4 P. Dehghanian, M. Fotuhi-Firuzabad, F. Aminifar, and R. Billinton, "A comprehensive scheme for reliability centered maintenance in power distribution systems—Part I: Methodology," *IEEE Trans. Power Deliv.*, vol. 28, no. 2, pp. 761–769, Apr. 2013.

5 P. Dehghanian, M. Fotuhi-Firuzabad, F. Aminifar, and R. Billinton, "A comprehensive scheme for reliability centered maintenance in power distribution systems—Part II: Numerical analysis," *IEEE Trans. Power Deliv.*, vol. 28, no. 2, pp. 771–778, Apr. 2013.

6 S. Bjarnadottir, Y. Li, and M. G. Stewart, "Hurricane risk assessment of power distribution poles considering impacts of a changing climate," *J. Infrastruct. Syst.*, vol. 19, no. 1, pp. 12–24, Apr. 2012.

7 A. Kwasinski, "Technology planning for electric power supply in critical events considering a bulk grid, backup power plants, and micro-grids," *IEEE Syst. J.*, vol. 4, no. 2, pp. 167–178, Jun. 2010.

8 M. H. J. Bollen, "Effects of adverse weather and aging on power system reliability," *IEEE Trans. Ind. Appl.*, vol. 37, no. 2, pp. 452–457, Mar. 2001.

9 P. Wang and R. Billinton, "Reliability cost/worth assessment of distribution systems incorporating time-varying weather conditions and restoration resources," *IEEE Trans. Power Deliv.*, vol. 17, no. 1, pp. 260–265, Jan. 2002.

10 M.A. Rios, D. S. Kirschen, D. Jayaweera, D. P. Nedic, and R. N. Allan, "Value of security: Modeling time-dependent phenomena and weather conditions," *IEEE Trans. Power Syst.*, vol. 17, no. 3, pp. 543–548, Aug. 2002.

11 Y. Zhou, A. Pahwa, and S. S. Yang, "Modeling weather-related failures of overhead distribution lines," *IEEE Trans. Power Syst.*, vol. 21, no. 4, pp. 1683–1690, Nov. 2006.

12 D. Bienstock and S. Mattia, "Using mixed-integer programming to solve power grid blackout problems," *Discret. Optim.*, vol. 4, no. 1, pp. 115–141, Mar. 2007.

13 J. Winkler, L. Duenas-Osorio, R. Stein, and D. Subramanian, "Performance assessment of topologically diverse power systems subjected to hurricane events," *Reliab. Eng. Syst. Saf.*, vol. 95, no. 4, pp. 323–336, Apr. 2010.

14 R. A. Francis, S. M. Falconi, R. Nateghi, and S. D. Guikema, "Probabilistic life cycle analysis model for evaluating electric power infrastructure risk mitigation investments," *Clim. Chang.*, vol. 106, no. 1, pp. 31–55, May 2011.

15 J. Tian and Y. Li, "System dynamics assessment of mitigation strategies for power distribution poles subjected to hurricanes," *Nat. Hazards*, vol. 70, no. 2, pp. 1263–1285, Jan. 2014.

16 S. K. Khator and L. C. Leung, "Power distribution planning: A review of models and issues," *IEEE Trans. Power Syst.*, vol. 12, no. 3, pp.1151–1159, Aug. 1997.

17 B. Ball, "Rebuilding electrical infrastructure along the Gulf coast: A case study," *Bridge-Washington-Nat. Acad. Eng.*, vol. 36, no. 1, pp. 21–26, Spring 2006.

18 N. Xu, S. D. Guikema, R. A. Davidson, L. K. Nozick, Z. Cagnan, and K. Vaziri, "Optimizing scheduling of post-earthquake electric power restoration tasks," *Earthquake Eng. Struc. Dynam.*, vol. 36, no. 2, pp. 265–284, Feb. 2007.

19 N. Perrier *et al.*, "A survey of models and algorithms for emergency response logistics in electric distribution systems. Part I: Reliability planning with fault considerations," *Comput. Oper. Res.*, vol. 40, no. 7, pp. 1895–1906, Jul. 2013.

20 N. Perrier *et al.*, "A survey of models and algorithms for emergency response logistics in electric distribution systems. Part II: Contingency planning level," *Comput. Oper. Res.*, vol. 40, no. 7, pp. 1907–1922, Jul. 2013.

21 J. H. Ellis, M. Jiang, and R. B. Corotis, "Inspection, maintenance, and repair with partial observability," *J. Infrastruct. Syst.*, vol. 1, no. 2, pp. 92–99, Jun. 1995.

22 R. B. Corotis, J. H. Ellis, and M. Jiang, "Modeling of risk-based inspection, maintenance and life-cycle cost with partially observable Markov decision processes," *Struct. Infrastruct. Eng.*, vol. 1, no. 1, pp. 75–84, Feb. 2005.

23 E. Byon, L. Ntaimo, and Y. Ding, "Optimal maintenance strategies for wind turbine systems under stochastic weather conditions," *IEEE Trans. Reliab.*, vol. 59, no. 2, pp. 393–404, Jun. 2010.

24 E. Byon and Y. Ding, "Season-dependent condition-based maintenance for a wind turbine using a partially observed Markov decision process," *IEEE Trans. Power Syst.*, vol. 25, no. 4, pp. 1823–1834, Nov. 2010.

25 N. L. Dehghani, Y. M. Darestani, and A. Shafieezadeh. "Optimal life-cycle resilience enhancement of aging power distribution systems: A MINLP-based preventive maintenance planning." *IEEE Access*, vol. 8, pp. 22324–22334, 2020.

26 I. Satkauskas, J. Maack, M. Reynolds, D. Sigler, K. Panda, and W. Jones, "Simulating impacts of extreme events on grids with high penetrations of wind power resources." *2022 IEEE/PES Transm. Distrib. Conf. Expos. (T&D)*, New Orleans, LA, USA, 2022, pp. 01–05, doi: https://doi.org/10.1109/TD43745.2022.9816884.

27 L. M. Maillart and L. Zheltova, "Structured maintenance policies on interior sample paths," *Nav. Res. Logist.*, vol. 54, no. 6, pp. 645–655, May 2007.

28 L. M. Maillart, "Maintenance policies for systems with condition monitoring and obvious failures," *IIE Trans.*, vol. 38, no. 6, pp. 463–475, Feb. 2006.

29 National Oceanic and Atmospheric Administration, 2014. [Online]. Available: http://www.noaa.gov (accessed December 2014).

30 Y. Lu and J. Garrido, "Doubly periodic non-homogeneous Poisson models for hurricane data," *Stat. Methodol.*, vol. 2, no. 1, pp. 17–35, Mar. 2005.

31 A. Arab, A. Khodaei, S. K. Khator, K. Ding, and Z. Han, "Post-hurricane transmission network outage management," in *Proc. IEEE Great Lakes Symp. Smart Grid New Ene. Eco.*, Chicago, IL, USA, Sep. 2013.

32 A. Arab, A. Khodaei, S. K. Khator, K. Ding, V. A. Emesih, and Z. Han, "Stochastic pre-hurricane restoration planning for electric power systems infrastructure," *IEEE Trans. Smart Grid*, vol. 6, no. 2, pp. 1046–1054, Jan. 2015.

33 A. Arab, E. Tekin, A. Khodaei, S. K. Khator, and Z. Han, "Dynamic maintenance scheduling for power systems incorporating hurricane effects," in *Proc. IEEE Smart Grid Comm. Conf.*, Venice, Italy, Nov. 2014.

34 Edison Electric Institute, "Before and after the storm: A compilation of recent studies, programs, and policies related to storm hardening and resiliency," Washington, DC, USA, Mar. 2014.

35 P. J. Vickery and L. A. Twisdale, "Prediction of hurricane wind speeds in the United States," *J. Struct. Eng.*, vol. 121, no. 11, pp. 1691–1699, Nov. 1995.

36 E. Wolfstetter, *Stochastic Dominance: Theory and Applications*. Berlin, Germany: Humboldt Univ., 1993.

37 R. J. Campbell, "Weather-related power outages and electric system resiliency," *Congressional Research Service*, Library of Congress, Aug. 2012.

38 S&C Electric Company's Products," Jul. 2014. [Online]. Available: http://www.sandc.com/products (accessed December 2014).

39 M. L. Puterman, *Markov Decision Processes: Discrete Stochastic Dynamic Programming*. New York, NY, USA: John Wiley & Sons, 1994.

40 P. Mandl, "Об асимптотическом поведении вероятностей в группах состояний однородной цепи маркова (in Russian)," *Časopis pro Pěstování Matematiky*, vol. 84, no. 2, pp. 140–149, 1959.

41 "IEEE guide for the interpretation of gases generated in oil-immersed transformers," IEEE Standard C57.104-2008, Feb. 2009.

42 J. D. McCalley *et al.*, "Automated integration of condition monitoring with an optimized maintenance scheduler for circuit breakers and power transformers," Final Project Rep., 2006. [Online]. Available: http://www.pserc.org (accessed December 2014).

4

The Economics of Capacity Planning for Climate Shocks

4.1 Principles of Resilience Investment

The link between climate change and heatwaves is well established in the literature. General Circulation Models (GCMs) of climate change estimate that the frequency and intensity of heatwaves will continue to increase, especially in the higher latitudes, affecting large areas worldwide that are not adequately adapted to this phenomenon. For instance, before the end of this century, Chicago is expected to experience a 25% increase in the frequency of heatwave events, while the annual heatwave days in Los Angeles will increase from 12 to up to 95 days [1]. Consequently, the electric power infrastructure is expected to be severely stressed by extreme heatwave events due to ever-increasing power demand, reduced transmission and generation capacity of the power system infrastructure, and increased risk of system failures, which can lead to large-scale blackouts and extended periods of power outages. On the other hand, the growing proliferation of grid-edge technologies and distributed, renewable resources in the energy mix to combat climate change introduces a new set of uncertainties to the system. The combination of these simultaneous developments calls for risk-informed strategic capacity planning and investments in resilience-building measures in the power infrastructure in the face of these events. In this chapter, we build upon our previous work to address this issue [2].

A heatwave can be defined as an extreme weather event in which, for at least two consecutive days/nights, the maximum/minimum temperature exceeds a certain high-percentile threshold, e.g., 90th and 99th percentiles of the temperature distribution [3]. Numerous studies in the literature investigated heatwaves from various perspectives. Authors in [3] used fragmentation metrics to investigate the spatiotemporal changes in the size and shape of heatwaves across North America. Reference [4] proposed a quantitative modeling and optimization framework for the design of power systems with operational flexibility and improved resilience

The Economics of Climate Resilience in Power Infrastructure, First Edition. Ali Arabnya and Amin Khodaei.

against extreme heatwave and drought events. Authors in [5] investigated the compound hazard risk of hurricanes, blackouts, and heatwaves and analyzed their variations due to climate change in Harris County, Texas as a case study. It was shown that moderate improvements in power distribution networks can significantly reduce the compound hazard risk. Their study suggests that climate adaptation strategies such as undergrounding the power distribution network and deploying distributed energy resources can significantly increase the resilience of coastal power systems against these extreme events. Authors in [6] proposed a machine learning model to predict power outages caused by heatwaves using historical failure data as well as operational and meteorological data for a large power distribution network in southern Italy. Reference [7] proposed a reliability-centered framework for identifying system upgrades as part of resilience-building measures in a medium voltage power distribution network against heatwaves. Reference [8] investigated the compound climate and infrastructure risk by analyzing the effects of electrical power grid failures on heatwave risk in three major metropolitan areas in the United States. They simulated the building-interior temperatures under concurrent heatwave and blackout events. It was concluded that the compound grid failure events and heatwaves of recent intensity and duration exposed between 68 and 100% of the urban population to a higher risk of heat exhaustion or heat stroke. Reference [9] developed a methodology to predict the risk of heatwaves and droughts to thermoelectric power plants using a set of integrated thermodynamic, regression, and climate models. It was shown that on average, a 1 °C rise in ambient cooling water temperature could cause a 0.15%–0.5% drop in power generation output. Reference [10] proposed a modeling framework to facilitate proactive planning by replacing power brownouts and blackouts with a rationing regime in various industrial sectors in Shanghai, China during heatwave events. The framework couples a supply-constrained input–output model with a model regressing daily electricity use and maximum temperature. The proposed model measures the magnitudes of gains and losses in each industrial sector based on total CO_2 emissions reduction, the economic cost of emissions reduction, and the total GDP loss for a given degree of electricity rationing enacted in the sector. The proposed model aimed at providing planners and policymakers with a tool to identify and choose decision alternatives based on their policy targets. Reference [11] examined the associations of heatwave and wildfire hazards with power outages and health outcomes using historical data in California cities. A regression analysis was conducted to estimate the power outage likelihood and the risk of emergency department visits and hospitalizations during hazard days. Results revealed that power outages were 2.24 times more likely, while outage hours were 18% higher during heatwave days in summer. In addition, emergency department visits and hospitalization rates for heat-related illnesses increased by over 1000% during joint heatwave outage events compared to about 800% during heatwave events without power outages. Authors in [12] presented an optimal operations management strategy for power systems under heatwave exposure. The proposed model incorporates the effects of hourly temperature on load profiles, power generation capacity in renewable and non-renewable resources, load curtailment, and transmission line capacity. The proposed operations aimed at quantifying the impact of heatwaves on transmission network operations. Reference [13] proposed a unified framework for the optimal allocation of available energy and demand-responsive resources in a power distribution system during heatwave events with the goal of minimizing operational costs and maintaining load-generation balance. The proposed model generates temperature set-points for the demand-responsive A/C units in order to regulate the demand. To prevent health risks to the residents, a conservative approach to utilize demand-responsive resources is adopted. A goal programming model that ensures the Pareto optimality of the solution was developed. The proposed model was extended in [14] for socially and environmentally aware power management under uncertainty in a residential neighborhood during heatwaves as a robust multi-objective problem with a different objective function and modeling approach for A/C power consumption. A comprehensive literature review of the dynamic thermal rating (DTR) in electrical power systems can be found in [15]. A systematic review of the state of the art in research, physical understanding, and scientific challenges of heatwaves can be found in [16]. Authors in [17] examined the sensitivity of power systems in the Western United States to climate extremes, such as heatwaves and droughts for a low and a moderate share of variable renewable energy in their energy mix. The study found that droughts increase the generation cost during summer. On the other hand, heatwaves have minimal impact on generation but increase Southern California's energy dependency. The authors highlight the importance of considering the risk of compounded climate events in planning and underscore the role of a robust interregional energy transfer capability in managing extreme events. Reference [18] investigated the impacts of reduced renewable energy generation and increased demand during heatwaves on system resilience. An operational model that utilizes long-term Hydrogen energy storage systems was introduced. The model aims to enhance system resilience by pre-charging the hydrogen energy storage before heatwaves and discharging it during these events.

An area that needs to be addressed in the literature is a forward-looking framework to proactively manage the risk of heatwaves in the transmission networks. The first step toward that goal is to quantify the risk of such events on the network. In this chapter, we propose a stress testing model to assess the risk of heatwaves on the capacity of power transmission networks to supply the demand under stressed conditions posed by these events. We aim to provide a strategic decision-making framework to equip system operators, utilities, and policymakers with a tool for operational planning, risk management, and capital allocation for resilience enhancement, considering the desired service level and underlying uncertainties associated with the heatwaves.

Considering the increasing frequency and intensity of such heatwaves, a data-driven, forward-looking framework is needed to proactively mitigate the risk of heatwaves in the transmission networks to meet the required service level under these extreme weather events. However, to the best of our knowledge, there is a lack of such models in the literature to enable capacity planning for transmission lines considering the required granularity level, as well as the need for modeling accuracy that incorporates the uncertainty associated with the heatwaves. That requires a modeling framework that simultaneously estimates the risk of future heatwaves in both power supply and demand. To address this gap, in this chapter, we propose a stochastic stress testing framework to assess the risk of heatwaves on the reduced capacity of power transmission networks in supplying the increased demand under conditions posed by these events. Our goal is to develop a tool that provides critical input to strategic decision-making by system operators, utilities, and policymakers for operational planning, risk management, and capital allocations to improve and maintain the system's resilience against heatwaves.

The remainder of this chapter is organized as follows. Section 4.2 describes the proposed model and associated techno-economic variables. Section 4.3 presents a model for the assessment of system capacity under stressed weather conditions. A case study along with numerical analysis is provided in Section 4.4, while Section 4.5 summarizes this chapter.

4.2 Climate Risk Assessment Variables

Consider a power transmission network in a geographical region under exposure to heatwave events. The rise in ambient temperature during heatwave events can result in a spike in power demand primarily due to air conditioning needs. On the other hand, the rise in the ambient temperature can lead to reduced power supply capacity due to the inverse relationship between the ampacity of transmission lines and ambient temperature. Thus, the transmission and distribution networks are expected to be simultaneously exposed to stress from both supply and demand sides during a heatwave event. The supply-side stress includes the reduction in both power generation and transmission and distribution network capacities. The reduction in power generation, however, is out of the scope of this study. The reduction in transmission and distribution networks includes a reduction in maximum power flow through lines and substations, plus failed lines and transformers due to overheating—which could trigger cascading outages and widespread blackouts. In this chapter, we focus on quantifying these risks on the capacity of transmission and distribution networks during such extreme weather events.

While heatwaves are generally defined as the exceedance of minimum and maximum daily temperatures above a certain high-percentile threshold for at least *two* consecutive days in a row, for the sake of practicality and without loss of generality, we decided to relax the duration clause in the above definition to any such events that would last for at least *one* day. The rationale for this modeling decision is due to the daily cyclical nature of the power system operations, which can be impacted by even one day of such extreme weather events. Therefore, to avoid underestimating system exposure to the risk of such events, we use the *day* as a unit of analysis for both power network operations and heatwaves. In addition, we use a 95th percentile threshold for both maximum and minimum daily temperatures in this study.

The arrival, duration, and intensity of heatwaves are stochastic in nature and can be modeled as stochastic processes. Hence, the stresses on both supply and demand due to a heatwave event are stochastic as well. The stochasticity of the power demand is attributed to the uncertainty in ambient temperature. The stochasticity of the energy supply, however, is due to the uncertainty associated with the combined impacts of ambient temperature, wind direction, and wind speed on the ampacity of the transmission lines. Therefore, a stochastic approach for stress testing the capacity of the transmission network is required to accurately quantify the system's vulnerability in meeting the demand during heatwave events. That insight on system performance during stressed conditions is critical for the utilities and system operators in making strategic investment decisions to maintain the desired service level considering the emerging climate risk landscape and extreme weather events.

4.3 An Economic Model for Resilience Investment

To formulate the problem, first, we model the heatwave arrival, duration, and magnitude using a set of stochastic processes and probabilistic models. Second, a model is developed to project the load profile for diurnal ambient temperatures. Third, we develop a stochastic DTR model for transmission lines to account for uncertainties in ambient temperature, wind speed, wind direction, and their impacts on the ampacity of power lines. Fourth, the problem is formulated as a stochastic mixed-integer program that minimizes load interruptions during heatwave scenarios. Finally, we leverage a well-established solution scheme to convert the stochastic mixed-integer program into its deterministic equivalent by using Monte Carlo sampling for scenario generation and a scenario reduction scheme to improve the computational efficiency of the model.

4.3.1 Heatwave Risk Modeling

Heatwave modeling includes the stochastic arrival rate of events, the duration of each event when it arrives, and the magnitude of the maximum and minimum daily temperatures when such events occur. Extreme value theory (EVT) is shown to be a powerful tool to model stochastic characteristics of extreme events, including heatwaves [19, 20]. The expected waiting time from a given heatwave event until the next event can be modeled as a point process occurring randomly in time with a fixed arrival rate. Without loss of generality, the occurrence of a heatwave event is modeled as a Poisson process, as follows [19]:

$$\Pr(m(\tau) = n) = \frac{(\lambda\tau)^n e^{-\lambda\tau}}{n!} \tag{4.1}$$

where

n is the number of heatwave events by day τ of the year
λ is the mean number of heatwave events per year

In an alternative modeling approach, a variable arrival rate can be used to account for interannual climate cycles in the geographical region where the power infrastructure is located.

EVT is a powerful methodology for modeling extreme values, such as significant deviations from the median of a probability distribution, with a wide range of applications in extreme climatological events. Therefore, to model the magnitude of the heatwaves, we use a Peaks-Over-Threshold (POT) approach that provides a modeling framework for independent exceedances of the maximum and minimum daily temperature above a threshold using a Generalized Pareto Distribution (GPD), as follows [20, 21]:

$$F(x) = 1 - \left[1 + \xi\left(\frac{x-\mu}{\sigma}\right)\right]^{-1/\xi}, \quad (x > \vartheta,\ \xi \neq 0) \tag{4.2}$$

where

ϑ is the temperature threshold
μ is the location parameter
σ is the scale parameter
ξ is the shape parameter for GPD distribution, assuming the exceedances are independent, identically distributed random variables

Since Eq. (4.2) is a cumulative distribution function, it can be rewritten as:

$$\Pr(X > x, x > \vartheta) = \Pr(x > \vartheta)[1 - F(x)] \tag{4.3}$$

$$\Pr(X > x) = \zeta_\vartheta\left[1 + \xi\left(\frac{x-\mu}{\sigma}\right)\right]^{-1/\xi} \tag{4.4}$$

where $\zeta_\vartheta = \Pr(x > \vartheta)$ is the probability of the exceedance of the temperature above a high threshold, ϑ.

This process can model the probability of exceedances of both daily maximum and minimum temperatures, T_{max} and T_{min}, above high thresholds ϑ_{max} and ϑ_{min}, respectively. While these thresholds are generally between 90th and 99th percentiles of their corresponding distributions [3], without loss of generality, a 95th percentile of the distributions for the maximum and minimum temperatures is adopted in this study.

To model the duration of the heatwaves—i.e., the length of successive days with upward crossings of the temperature threshold—we can use an exponential-like distribution, such as a discrete geometric distribution, as follows [19, 22]:

$$\Pr(HD) = (1 - \theta)^{HD-1}\theta \tag{4.5}$$

where

HD is the duration of the heatwave event
θ is the reciprocal of the mean duration.

4.3.2 Load Forecasting

To model the effects of temperature rise on power demand in the system, the sensitivity of the load profiles for each class of customers (i.e., residential, industrial, commercial, and critical) to temperature rise can be calculated and aggregated. The increase of power demand at bus b due to temperature rise is modeled using a Simple Linear Regression (SLR) model, as follows [23]:

$$\Delta D_{bt} = TS_{bt} D_{bt}^{mean} \frac{\Delta T_t}{T_t^{mean}} \tag{4.6}$$

where

ΔD_{bt} is the percentage increase in power consumption at bus b for each degree increase in ambient temperature T_t at time t compared to mean temperature T_t^{mean}
TS_{bt} is the temperature sensitivity of load D_{bt} in bus b at time t of the day.

As shown, the ambient temperature at any given time of the day is the independent variable in Eq. (4.6) which needs to be determined. However, predicting the daily temperature pattern is a challenging task involving modeling uncertainty. To address this issue, we use the random variables for the maximum and the minimum daily temperatures (derived in the previous subsection) and develop a daily temperature profile using the methods presented in [24]. The ambient temperature during the daytime, i.e., the hours between the *apparent* sunrise and sunset, is obtained using the sinusoidal relationship between time of the day and daily temperature extrema, as follows:

$$T_t = T^{min} - \frac{T^k}{2} + \frac{1}{2}\sqrt{(T^k)^2 + 4(T^{max} - T^{min})\left(1 + \frac{T^{max} - T^{min}}{T^k}\right) T^k S_t} \tag{4.7}$$

$$S_t = \sin\left(\pi \frac{t - LSH + (DL/2)}{DL + 2Y}\right), \quad \forall t = t_r, t_r + 1, \dots, t_s \tag{4.8}$$

where

T^k is a parameter that specifies the temperature increment that sensible heat flux is doubled compared to a situation without buoyancy
LSH is the time of maximum solar height
DL is the day length
Y is the delay between T^{max} and LSH.

The ambient temperature during the night, i.e., from apparent sunset to apparent sunrise the next day, is modeled using the exponentially declining relationship between time and temperature during those hours, as follows:

$$T_t = \frac{\left(T_{ND}^{min} - T_{t_s} \exp\left(-\frac{NL}{h}\right) + \left(T_{t_s} - T_{ND}^{min}\right) \exp\left(-\frac{t-t_s}{h}\right)\right)}{\left(1 - \exp\left(-\frac{NL}{h}\right)\right)},$$
$$\forall t = t_s, t_s + 1, \dots, t_r - 1, t_r \tag{4.9}$$

where

T_{ND}^{min} is the minimum temperature on the next day
NL is the night length (i.e., 24–DL)
t_s is the time of sunset
h is a time coefficient.

4.3.3 Stochastic Dynamic Thermal Rating

The surface temperature of a transmission line is a function of its material properties, surface condition, diameter, electrical current flown through the conductor, and ambient weather conditions. To ensure a safe and reliable operation in a transmission network, its transmission lines' surface temperature must remain below a certain threshold specified by the manufacturer [25]. Therefore, the maximum capacity of a transmission line can proportionally be constrained by the surface temperature threshold specified by its manufacturer. DTR incorporates weather-related factors such as wind direction, wind speed, and ambient temperature to compute the ampacity of transmission lines. IEEE Standard 738-2006 defines the steady-state heat balance for calculating the current temperature of bare overhead conductors as follows [25]:

$$A_l(T_t) = \sqrt{\frac{Q_c\left(T_l^c, T_t, V_s, \phi_s\right) + Q_r\left(T_l^c, T_t\right) - Q_s}{R\left(T_l^c\right)}} \tag{4.10}$$

where

Q_c is convected heat loss rate per unit length
Q_r is radiated heat loss rate per unit length
Q_s is heat gain rate from the sun
$R\left(T_l^c\right)$ is conductor resistance at critical temperature T_l^c
V_s and ϕ_s are the wind speed and direction in weather scenario s.

To account for the stochasticity of wind speed and direction for a given ambient temperature T_t, we replace wind speed V_s and wind direction ϕ_s variables with two independent random variables V_s and ϕ_s, respectively. Without loss of generality, the stochasticity of the wind speed can be modeled with a lognormally distributed random variable $V_s \sim Lognormal(\mu_V, \sigma_V)$, as shown in our previous study [26]. The stochastic wind direction, which takes a value in the range of $[0, 2\pi]$ can be modeled using a wide range of circular distributions. Without loss of generality, we use a von Mises distribution as it was used in [27] to model the stochastic wind direction, as follows:

$$f(\phi_s \mid \mu_\phi, \kappa) = \frac{e^{\kappa \cos(\phi_s - \mu_\phi)}}{2\pi I_0(\kappa)} \tag{4.11}$$

where

$I_{0(\kappa)}$ is a modified Bessel function of order 0
μ_ϕ is a measure of location
κ is a measure of concentration.

With this replacement, the maximum capacity of transmission line l under ambient temperature T_t at time t in weather scenario s becomes a random variable denoted by $PL_{lTts^{max}}$. The convected heat loss rate, which is a function of wind speed, wind direction, and ambient temperature, is obtained as follows:

$$Q_c = max(Q_{cn}, Q_{c1}, Q_{c2}) \tag{4.12}$$

$$Q_{cn} = 0.0205\, \rho_f^{0.5}\, d^{0.75}\left(T_l^c - T_t\right)^{1.25} \tag{4.13}$$

$$Q_{c1} = \left(1.01 + 0.0372\left(\frac{d\rho_f V_s}{\gamma_f}\right)^{0.52}\right) k_f K_s (T_{cl} - T_t) \tag{4.14}$$

$$Q_{c2} = 0.0119\left(\frac{d\rho_f V_s}{\mu_f}\right)^{0.6} k_f K_s \left(T_l^c - T_t\right) \tag{4.15}$$

$$K_s = 1.194 - \cos(\phi_s) + 0.194\cos(2\phi_s) + 0.368\sin(2\phi_s) \tag{4.16}$$

where

Q_{cn} is the convected heat loss when there is no wind
Q_{c1} is the convected heat loss at low wind
Q_{c2} is the convected heat loss at high wind
ρ_f is the air density
d is the conductor diameter
γ_f is the dynamic viscosity of the air
k_f is the thermal conductivity of the air
K_s is the wind direction factor.

The radiated heat loss—which is a function of the ambient temperature—is derived as follows:

$$Q_r = 0.0178d\varepsilon\left(\left(\frac{T_l^c + 273}{100}\right)^4 - \left(\frac{T_t + 273}{100}\right)^4\right) \tag{4.17}$$

where ε is the coefficient of emissivity. Finally, the rate of solar heat gain is calculated, as follows:

$$Q_s = \alpha Q_{se}\sin(\arccos(\cos(H_c)\cos(Z_c - Z_l)))A \tag{4.18}$$

where

α is the solar absorptivity
Q_{se} is the total heat flux rate elevation correction factor
H_c is the altitude of the sun
Z_c is the azimuth of the sun
Z_l is the azimuth of line l
A is the projected area of conductor per unit length.

4.3.4 Stochastic Power Dispatch

The power dispatching operations during a heatwave event are subject to uncertainties associated with the power demand, the ampacity of transmission lines, and the characteristics of the heatwave events. Given the crucial importance of maintaining a maximum service level in supplying the power demand during a heatwave, the economic considerations on unit commitment and power dispatch become a secondary priority for utilities and system operators. Therefore, we formulate the power dispatch problem under the heatwave event as a stochastic mixed-integer program with an objective function that minimizes the load interruption to the extent feasible, given the physical constraints of the power system, as:

$$\min_{LI} \sum_{t} \sum_{b} \sum_{s} p(s) VOLL_{bt} \, LI_{bts} \tag{4.19}$$

s.t.

$$\sum_{i \in N_b} P_{its} + \sum_{l \in N_b} PL_{lts} - LI_{bts} = D_{bts}, \quad \forall b, \forall t, \forall s \tag{4.20}$$

$$P_i^{min} I_{its} \leq P_{its} \leq P_i^{max} I_{its}, \quad \forall i, \forall t, \forall s \tag{4.21}$$

$$-PL_{lT_ts}^{max} \leq PL_{lts} \leq PL_{lT_ts}^{max}, \quad \forall l, \forall t, \forall s, \forall T_t \tag{4.22}$$

$$PL_{lts} = \frac{\sum_b \beta_{lb} \delta_{bts}}{x_l}, \quad \forall l, \forall b, \forall t, \forall s \tag{4.23}$$

where

$p(s)$ is the probability of scenario s
$VOLL_{bt}$ is the value of lost load in bus b at time t
LI_{bts} is the load interruption variable in bus b at time t under scenario s.

Equation (4.20) represents the load balance constraints where P_{its} is the real power generation variable of unit i at time t under scenario s, PL_{lts} is power flow variable for line l at time t under scenario s, and D_{bts} is load demand at bus b at time t in scenario s. This constraint ensures the injected power to a bus from connected transmission lines and generation units supply the bus load in each scenario; however, if the injected power is not sufficient, the load supply will be interrupted equal to the load interruption variable LI_{bts}. Equation (4.21) represents the power generation constraints where I_{its} is a binary unit commitment variable, and P_i^{min} P_i^{max}, respectively, are the minimum and maximum power generation capacity in generation unit i. Equations (4.22) and (4.23) represent the power flow constraints where $PL_{lT_ts}^{max}$ is the maximum power flow capacity of line l at temperature T_t under scenario s, β_{lb} is an element of line l and bus b in line-bus incidence matrix, and δ_{bts} is the bus voltage angle.

4.3.5 Solution Scheme

Due to the presence of continuous dependent random variables in the model, i.e., power demand and transmission line ampacity, the stochastic data process of the proposed models, ψ, has infinite support. To make the problem computationally tractable, the stochastic data process ψ needs to be redistributed to provide finite support with a reduced number of representative scenarios. To do so, we use a well-established solution scheme that we previously utilized in [28].

We use Monte Carlo sampling [29] to replace ψ by a scenario tree approximation ψ_S which has a finite but large number of scenarios. An adequately large number of samples are generated to ensure that the whole range of values for a multivariate random variable is sampled. For a sample size N, each sample includes four random values, including minimum daily temperature, maximum daily temperature, wind direction, and wind speed from their corresponding distributions. Therefore, N scenarios are constructed, each with a probability of $1/N$. In the next step, we reduce the number of scenarios to a computationally tractable size. Various reduction techniques are available for different applications. For the constructed probability measure of ψ_S, it is required to determine an index set $K_* \subset \{1, \ldots, N\}$ of a given cardinality $\#K'_* = N - N'$ and probability measure ψ^* such that,

$$\begin{aligned} \widehat{\mu}_c(\psi_S, \widetilde{\psi}_*) = \inf \Bigg\{ \widehat{\mu}_c \left(\psi_S, \sum_{k=1, k \neq K'_*}^{N} p_{k'} s_{k'} \right) : K \subset \{1, \ldots, N\}, \#K'_* \\ = N - N' \sum_{k' \notin K'_*} p_{k'} = 1, \quad p_{k'} \geq 0, \ k' \notin K' \Bigg\} \end{aligned} \tag{4.24}$$

where the Kantorovich function $\widehat{\mu}_c(\psi_S, \widetilde{\psi}_*)$ is an estimation of the probability distance $\zeta_c(\psi_S, \widetilde{\psi}_*)$.

Problem (4.24) can be solved through various techniques, but due to the accuracy of the backward reduction algorithm, we solve it through this method. Readers are referred to [30] for a detailed explanation of the backward scenario reduction algorithm.

By constructing a scenario tree and reducing it to a tractable number of scenarios, the proposed stochastic program is converted to its deterministic mixed-integer program equivalence that is computationally efficient.

4.4 A Heatwave Case Study

As a case study, we analyze the IEEE 118-bus system to illustrate the application and effectiveness of the proposed model. The IEEE 118-bus test case system represents a portion of the electric power network in the Midwestern United States in the 1960s. However, this system does not include all the required data for our analysis in this study, including physical characteristics of conductors, load sensitivity to temperature, weather, and geographical data of the system. To address that gap, we use the weather and geographical data in Chicago, IL, as a representative city in the American Midwest to ensure the compatibility of the test system with the weather and geographical information.

We obtained the weather data from the weather station in Chicago Midway International Airport (MDW) from January 1990 through December 2020 via two different open data sources. The hourly wind data, including wind speed and direction in MDW during this timeframe were obtained through Iowa Environmental Mesonet (IEM) [31]. The hourly temperature data were obtained from NOAA's National Centers for Environmental Information (NCEI) [32]. The solar position data, including the sunrise and sunset on various days of the year in Chicago, were obtained from NOAA's Earth System Research Laboratories [33]. For illustration, we assumed an Aluminum Conductor Steel Reinforced (ACSR) cable for all transmission lines in the IEEE 118-bus system across the board and used the corresponding data provided in [25]. Due to a lack of detailed topological data on transmission lines in the IEEE 118-bus system—and considering the overall direction of the entire system map that is approximately extended from northwest to southeast ordinal angles—for the sake of illustration, we assumed an azimuth of 135° for all transmission lines across the board. Chicago's latitude and elevation above the sea level (obtained from Google Earth), i.e., 41°N and 190 m, respectively, were used for the entire system.

We considered the loads at bus *B4* as critical (e.g., medical centers or water treatment plants), *B11* as commercial, and *B15* as industrial loads. The rest of the loads in the system were considered residential loads. From our previous study in [34], the value of lost load is considered $3.706/kWh for industrial, $6.979/kWh for commercial, and $0.110/kWh for residential loads. However, for the critical load, we assume a value of $10/kWh to signify higher importance. We use the weather data during the warm and hot seasons, from April 1st through September 30th of each year from 1990 to 2020. Using Python and through maximum likelihood estimation (MLE) method, we estimated the parameters for corresponding distribution functions (described in Section 4.3) for the daily maximum and minimum temperature exceedances, as well as wind speed and directions (i.e., two separate GPDs for daily maximum and minimum exceedances, a Poisson distribution for heatwave arrival process, a geometric distribution for a heatwave duration,

Table 4.1 Parameters of fitted probability distribution functions.

Random variable	Distribution	Parameter	Estimated value
Max. temperature	Generalized pareto	σ	4.082
		ξ	−0.210
		ϑ_{max}	32.2
Min. temperature	Generalized pareto	σ	3.723
		ξ	−0.245
		ϑ_{min}	21.1
Heatwave arrival	Poisson	λ	11.483
Heatwave duration	Geometric	HD	1.854
Wind speed	Lognormal	μ_V	1.773
		σ_V	1.182
Wind direction	von Mises	μ_ϕ	−0.007
		κ	0.048

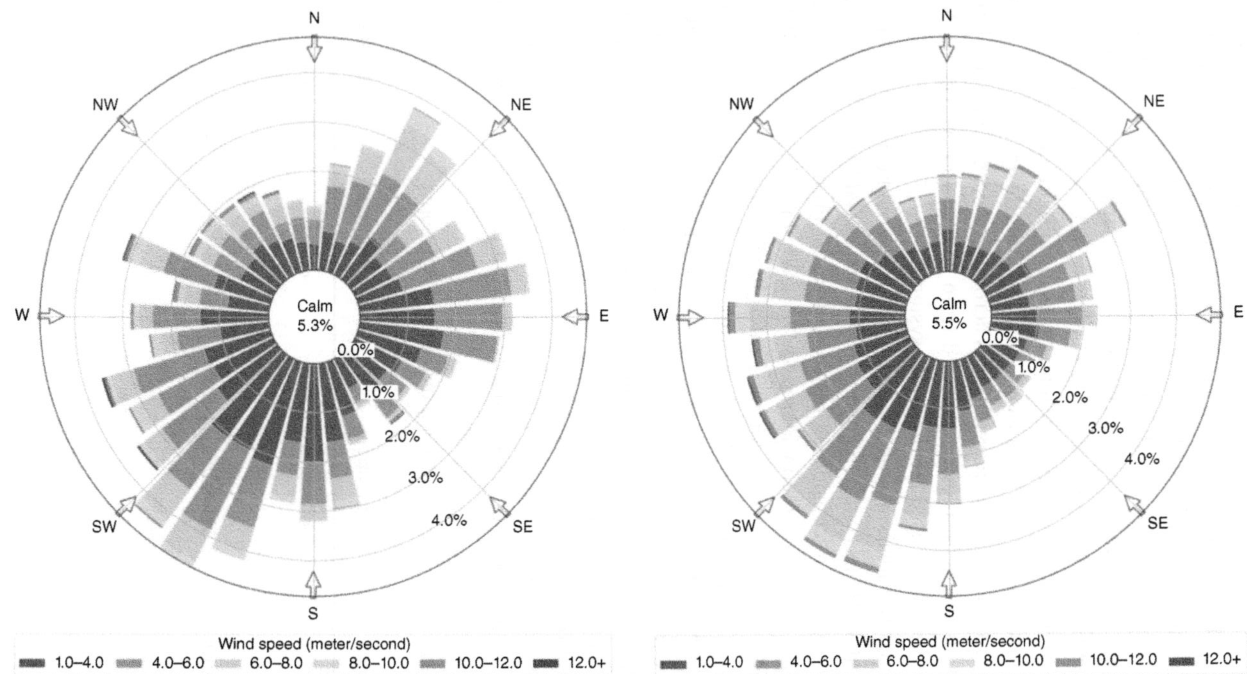

Figure 4.1 Wind roses of April–September (left) versus the entire year (right).

a lognormal distribution for wind speed, and a von Mises distribution for wind direction). A set of standard quantitative and qualitative statistical diagnostic tests were performed to ensure statistical soundness of all fitted distributions. Table 4.1 summarizes the derived parameters for each of these probability distributions. The wind roses in Figure 4.1 visualize the dynamics of the wind behavior during the warm season compared to its behavior during the year. To illustrate the model, we analyze three heatwave scenarios, as follows:

Baseline scenario: A stochastic set of heatwave events that represent a long-run average scenario is considered. In this scenario, stochastic wind speed and direction are included.

Adverse scenario: A heatwave event with a 10-year return period is considered. A zero wind speed is assumed.

Severely adverse scenario: A heatwave event with a 100-year return period is considered. Similar to the Adverse Scenario, a zero wind speed was assumed to study the system's behavior under an extremely stressed weather condition.

Given the load profile in the IEEE 118-bus system, this system has a relatively large transmission reserve margin. To illustrate the dynamics of the proposed model, we use two cases for the load profile in each scenario, as follows:

Case I: The existing load profile in the IEEE 118-bus system is used in our analysis.

Case II: Through an auxiliary mixed-integer program, we modify the existing load profile in the IEEE 118-bus system by increasing them up to a level that, under normal weather conditions, still all transmission lines maintain a reserve margin of 15% or higher. In that modification, we multiplied all loads by a factor of 1.581.

Table 4.2 Reduced scenarios for the baseline.

s	$p(s)$	T^{max} (°C)	T^{min} (°C)	V_s (m/s)	ϕ_s (rad)
1	0.108	33.2	22.1	1.2	1.52
2	0.071	36.1	23.5	6.7	−1.39
3	0.065	32.9	22.4	4.6	0.30
4	0.118	32.6	25.3	2.7	−1.02
5	0.169	34.1	23.0	0.8	−1.55
6	0.034	34.4	23.4	31.4	1.26
7	0.085	33.0	21.9	0.9	−2.23
8	0.172	33.4	23.4	2.2	0.61
9	0.098	33.3	23.9	3.6	2.61
10	0.08	38.1	22.1	4.5	0.17

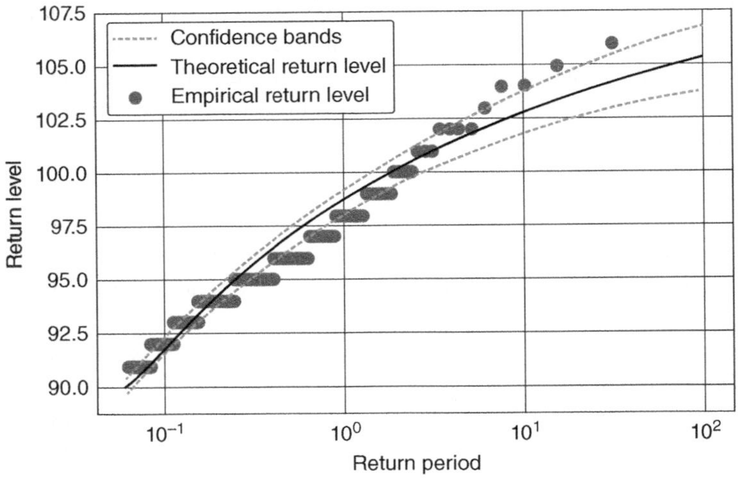

Figure 4.2 Return period plot (years) for maximum temperature exceedance (°F).

For Baseline Scenario, we generated 10,000 scenarios using the Monte Carlo sampling method. Using a backward reduction algorithm, the number of scenarios was reduced to 10 representative scenarios to improve the computational tractability of the stochastic mixed-integer program described in (4.19)–(4.23). Table 4.2 shows the representative scenarios after scenario reduction with their associated probabilities. For Adverse and Severely Adverse Scenarios, we use the return levels of maximum and minimum daily temperature exceedances for 10-year and 100-year return periods, respectively. Figure 4.2 illustrates the return period plot of the maximum daily temperature exceedance (in °F). A 100-year heatwave event can raise the maximum daily temperature in Chicago to 105.3 ± 1.5 °F (i.e., up to 41.5 °C) on a given day. The return levels for minimum daily temperature in an event with a 100-year return period were estimated at 83.7 ± 1.0 °F (i.e., up to 29.3 °C). The maximum and minimum daily temperatures in a 10-year heatwave event were estimated at 103 °F (39.4 °C) and 82 °F (27.7 °C), respectively.

Using the derived maximum and minimum daily temperatures for each of these scenarios and through (4.7)–(4.9), we develop a daily temperature profile. In the next step, the demand increases with respect to base load profiles in Cases I and II—which are based on average daily temperature from April to September—were calculated. Next, the ampacity of transmission lines is calculated using the projected diurnal temperature profiles for each scenario. Finally, we implemented the mixed-integer program (4.19)–(4.23) in GAMS and solved it using IBM's CPLEX solver. As illustrated in Figure 4.3, the expected load interruptions in Cases I and

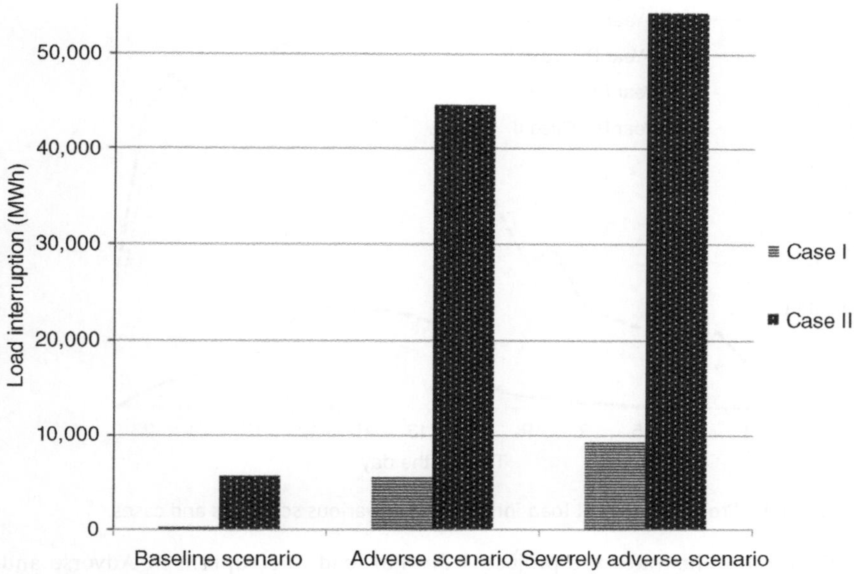

Figure 4.3 Expected load interruption under various hazard scenarios and cases.

II of the Baseline Scenario are estimated at 0.77 and 5704 MWh/day, respectively. The expected load interruption in Cases I and II of Adverse and Severely Adverse Scenarios is significantly higher than in the Baseline Scenario. The expected value of the heatwave duration is estimated at 1.8 days, while the expected value of heatwave events (with various magnitudes) was estimated at 11.4 days/year. It is worth noting that while a 100-year event has only a 1% chance of occurrence each year, they are not as rare as they may seem: statistically, the probability of observing at least one such event in a decade is slightly less than 10%.

The daily cost of load interruption, considering the value of lost load in (4.19), is estimated at $85 and $62,420 for Case I and Case II of the Baseline Scenario, respectively. In the Adverse Scenario, the daily cost of load interruption significantly rises and is estimated at $598,836 and $9,271,693 in Cases I and II, respectively. Finally, in the Severely Adverse Scenario, the daily cost of load interruption increases even further and is estimated at $1,234,019 and $12,216,730 in Cases I and II, respectively. It is worth noting that these cost figures do not represent the direct cost imposed on the utility. Rather, they represent the opportunity cost of interruptions from the consumer perspective as the value of lost load reflects the total average price per unit consumers are willing to pay to avoid the interruption.

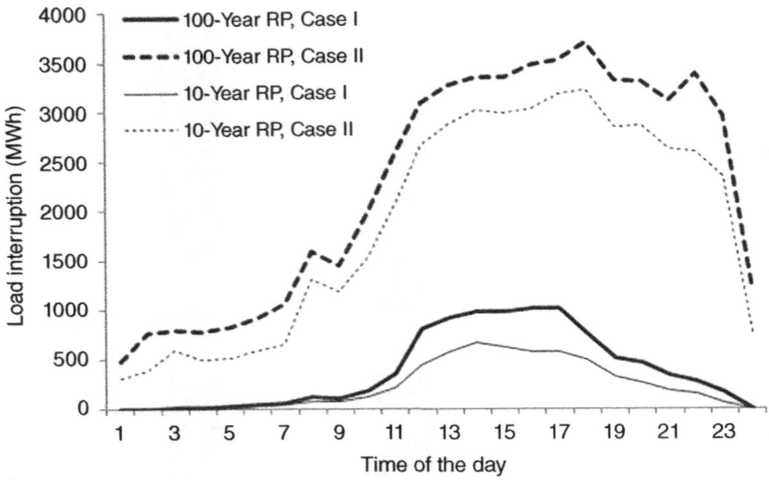

Figure 4.4 Projected diurnal load interruption in various scenarios and cases.

Figure 4.4 illustrates the projected diurnal load interruption in Adverse and Severely Adverse Scenarios. As expected, a 100-year heatwave event results in a higher load interruption than a 10-year event. However, Case II of both scenarios leads to a significantly higher level of load interruption in the system than Case I. That is due to a lower level of transmission reserve in the system under Case I. This demonstrates the importance of maintaining an adequate transmission reserve as a buffer against future heatwave events.

4.5 Summary

In this chapter, we proposed a stochastic stress testing framework to assess the risk of heatwave events on the operational capability of the transmission network in meeting the increased power demand in conjunction with a reduced transmission capacity of the system during an extreme temperature episode. It was shown that extreme heatwave events could render the transmission capacity to a dangerously low level, leading to incapacitating the system in supplying the much-needed power demand during such events. Our analysis shows that maintaining a risk-informed, adequate level of transmission reserve can significantly hedge the risk of such events in power systems. The proposed framework can be effectively used by utilities, system operators, urban planners, and policymakers as a strategic decision-making tool in managing the load profiles, transmission reserve, and power system expansion decisions in the face of heatwaves and extreme temperature conditions.

References

1 G. Luber and M. McGeehin, "Climate change and extreme heat events," *Am. J. Prev. Med.*, vol. 35, no. 5, pp. 429–435, 2008.

2 A. Arab, A. Khodaei, and S. K. Khator, "Stochastic dynamic thermal rating of transmission lines in heatwave events," in *Proc. IEEE Innovative Smart Grid Technol. (ISGT) Europe*, Novi Sad, Serbia, Oct. 2022, pp. 1–5.

3 D. Keellings, E. Bunting, and J. Engström, "Spatiotemporal changes in the size and shape of heat waves over North America," *Clim. Chang.*, vol. 147, pp. 165–178, 2018.

4 I. F. Abdin, Y. P. Fang, and E. Zio, "A modeling and optimization framework for power systems design with operational flexibility and resilience against extreme heat waves and drought events," *Renew. Sust. Energ. Rev.*, vol. 112, pp. 706–719, 2019.

5 K. Feng, M. Ouyang, and N. Lin, "Tropical cyclone-blackout-heatwave compound hazard resilience in a changing climate," *Nat. Commun.*, vol. 13, no. 1, pp. 4421, 2022.

6 M. Atrigna *et al.*, "A machine learning approach to fault prediction of power distribution grids under heatwaves," *IEEE Trans. Ind. Appl.*, vol. 59, pp. 4835–4845, 2023.

7 L. Bellani *et al.*, "A reliability-centered methodology for identifying renovation actions for improving resilience against heat waves in power distribution grids," *Int. J. Electr. Power Energy Syst.*, vol. 137, pp. 107813, 2022.

8 B. Stone *et al.*, "Compound climate and infrastructure events: how electrical grid failure alters heat wave risk," *Environ. Sci. Technol.*, vol. 55, no. 10, pp. 6957–6964, 2021.

9 M. A. Cook, C. W. King, F. T. Davidson, and M. E. Webber, "Assessing the impacts of droughts and heat waves at thermoelectric power plants in the United States using integrated regression, thermodynamic, and climate models," *Energy Rep.*, vol. 1, pp. 193–203. 2015.

10 Z. Liang *et al.*, "Heat wave, electricity rationing, and trade-offs between environmental gains and economic losses: The example of Shanghai," *Appl. Energy*, vol. 184, pp. 951–959, 2016.

11 K. Loa, "Ovens and tinderboxes in the golden state: A mixed-methods exploratory analysis of joint heatwave power outage events," *The Pardee RAND Graduate School*, Doctoral dissertation, 2022.

12 A. Alawad, A. Alnakhli, and P. Dehghanian, "Optimal energy management of a power transmission grid under a heatwave exposure," in *Proc. North American Power Symp.*, College Station, TX, USA, 2021, pp. 1–6.

13 M. Choobineh, A. Speake, M. Harris, P.C. Tabares-Velasco, and S. Mohagheghi, "End-user-aware community energy management in a distribution system

exposed to extreme temperatures," *IEEE Trans. Smart Grid*, vol. 10, no. 4, pp. 3753–3764, 2018.

14 A. Dindar, "Socially-and-environmentally-aware power management in residential neighborhoods exposed to heat waves considering uncertainties," Master's thesis, Colorado School of Mines, Golden, CO, USA, 2020.

15 C.M. Lai, and J. Teh, "Comprehensive review of the dynamic thermal rating system for sustainable electrical power systems," *Energy Rep.*, vol. 8, pp. 3263–3288, 2022.

16 D. Barriopedro, R. García-Herrera, C. Ordóñez, D. G. Miralles, and S. Salcedo-Sanz, "Heat waves: Physical understanding and scientific challenges," *Rev. Geophys.*, vol. 61, no. 2, pp. 1–54, 2023.

17 A. Dyreson *et al.*, "The role of regional connections in planning for future power system operations under climate extremes," *Earth's Future*, vol. 10, no. 6, 2022, Art No. e2021EF002554.

18 W. Yin and Y. Hou, "Resilient operation of long-term hydrogen energy storage-assisted power systems against heatwave events," in *2023 IEEE PES Innovative Smart Grid Technologies Europe (ISGT EUROPE)*, IEEE, 2023, pp. 1–5.

19 D. Keellings and P. Waylen. "Increased risk of heat waves in Florida: Characterizing changes in bivariate heat wave risk using extreme value analysis," *Appl. Geogr.*, vol. 46, pp. 90–97, Jan. 2014.

20 S. Coles, *An Introduction to Statistical Modeling of Extreme Values*, London, UK: Springer, 2001.

21 M. Tanarhte, P. Hadjinicolaou, and J. Lelieveld, "Heat wave characteristics in the eastern Mediterranean and Middle East using extreme value theory," *Climate Res.*, vol. 63, no. 2, pp. 99–113, Apr. 2015.

22 D. Wilks, *Statistical Methods in the Atmospheric Sciences*. Amsterdam, The Netherlands: Elsevier, 2019.

23 C. S. Chen, M. S. Kang, J. C. Hwang, and C. W. Huang, "Temperature effect to distribution system load profiles and feeder losses," *IEEE Trans. Power Syst.*, vol. 16, no. 4, pp. 916–921, Nov. 2001.

24 J. E. Ephrath, J. Goudriaan, and A. Marani, "Modelling diurnal patterns of air temperature, radiation wind speed and relative humidity by equations from daily characteristics," *Agric. Syst.*, vol. 51, no. 4, pp. 377–393, Aug. 1996.

25 D. Douglass and J. Reding, "IEEE standard for calculating the current-temperature of bare overhead conductors," IEEE Standard 738–2006, Jan. 2007.

26 A. Arab, E. Tekin, A. Khodaei, S. K. Khator, and Z. Han, "System hardening and condition-based maintenance for electric power infrastructure under hurricane effects," *IEEE Trans. Reliab.*, vol. 65, no. 3, pp. 1457–1470, Sep. 2016.

27 N. Masseran, A. M. Razali, K. Ibrahim, and M. T. Latif, "Fitting a mixture of von Mises distributions in order to model data on wind direction in Peninsular Malaysia," *Energy Convers. Manag.*, vol. 72, pp. 94–102, Aug. 2013.

28 A. Arab, A. Khodaei, S. K. Khator, K. Ding, V. A. Emesih, and Z. Han, "Stochastic pre-hurricane restoration planning for electric power systems infrastructure," *IEEE Trans. Smart Grid*, vol. 6, no. 2, pp. 1046–1054, Mar. 2015.

29 M. D. McKay, R. J. Beckman, and W. J. Conover, "Comparison of three methods for selecting values of input variables in the analysis of output from a computer code," *Technometrics*, vol. 21, no. 2, pp. 239–245, 1979.

30 H. Heitsch and W. Rmisch, "Scenario reduction algorithms in stochastic programming," *Comput. Optim. Appl.*, vol. 24, nos. 2–3, pp. 187–206, 2003.

31 The Iowa Environmental Mesonet (IEM). University of Iowa. https://mesonet.agron.iastate.edu (accessed Dec. 2021).

32 National Centers for Environmental Information (NCEI). https://www.ncei.noaa.gov (accessed Apr. 2021).

33 Global Monitoring Laboratory. The NOAA Earth System Research Laboratories. https://gml.noaa.gov (accessed Dec. 2021).

34 A. Arab, A. Khodaei, S. K. Khator, and Z. Han, "Electric power grid restoration considering disaster economics," *IEEE Access*, vol. 4, pp. 639–649, Feb. 2016.

5

The Economics of Resilience Planning in Power Infrastructure Expansion

5.1 Background

Long-term resilience planning for power infrastructure involves optimal capacity expansion, considering the risk of disruptions and contingencies. Power system planning is the science of determining the optimal place, size, and time for adding new resources to power systems. Previous studies investigated generation and transmission expansion planning methodologies in competitive market environments. The most common planning methods are based on mathematical optimization, such as branch and bound, linear programming, and Benders decomposition [1–9]. Also, heuristic approaches, such as genetic algorithms, fuzzy sets, simulated annealing, expert systems, and game-theoretic methods, were used to solve the expansion planning problem [10–18]. Market-based planning in power systems considers economic, security, and reliability constraints and analyzes the risk of planning strategies based on uncertainties [19–23]. Using Lagrangian relaxation and Benders decomposition, the impact of transmission security on generation resource planning was discussed [24]. The coordination of transmission expansion planning with competitive generation capacity planning was presented in [25]. In [26], a proposal was offered on the interaction of generation and transmission investments in transmission planning. In [27], the long-term transmission expansion planning problem in a competitive pool-based electricity market was modeled. In this model, several scenarios based on the future system demand were defined, and the optimal expansion planning was simulated by maximizing the aggregated social welfare. A bilevel model for transmission expansion planning within a market environment was proposed in [28]. The minimization of the network investment was considered at the upper level, while the lower level included pool trading constraints.

Authors in [29] analyzed the incorporation of wind power into modern electricity markets using a unique stochastic bilevel mixed integer optimization model.

The Economics of Climate Resilience in Power Infrastructure, First Edition. Ali Arabnya and Amin Khodaei.

The model includes binary variables in the lower level to represent line-switching decisions, thereby improving transmission capability under varying load profiles. To solve the complex bilevel mixed integer program, the authors customize a decomposition method and develop grid structure-based enhancement techniques. The numerical results show that topology control (TC) significantly reduces wind energy curtailment and increases wind penetration levels. Ref. [30] investigated the limitations of traditional transmission expansion planning and proposed the integration of Seasonal Network Optimization (SNO) to address these limitations. Unlike the static approach of conventional transmission expansion planning models, SNO can dynamically adjust the network configuration to accommodate seasonal variations in load profiles. The authors presented two enhanced models: The first model optimizes the status of existing transmission lines, while the second model optimizes both existing and new lines. This dynamic approach can help facilitate a more efficient and cost-effective power expansion process capable of meeting future demand growth while adapting to seasonal load variations.

Authors in [31] proposed a reduced disjunctive model for transmission expansion planning designed to enhance the flexibility and capacity of power systems in accommodating wind power. The proposed model combines the unit commitment and the optimal transmission switching to handle the uncertainties and fluctuations associated with wind energy. By reducing the number of variables compared to a traditional disjunctive model, the computational efficiency is improved. Ref. [32] introduced an advanced reconfiguration strategy for microgrids in active distribution networks. A risk-averse, two-stage mixed integer conic program model was proposed to support planning for networked microgrids, incorporating capacity expansion and seasonal reconfiguration decisions. The model is validated under stochastic islanding scenarios using a second-order conic program, with a conditional value-at-risk (CVaR) measure to control islanding risks. It was shown that the model improves cost-efficiency and system performance during both normal operations and emergency islanding conditions. In [33] a transmission and energy storage expansion planning model designed to handle extreme events was proposed. The proposed model considers three types of uncertainties including component failure rates, likelihood of contingencies in different zones, and total failed components, represented by interval, box, and N-k uncertainty sets, respectively. The developed min-max-min expansion planning model aims to optimize expansion plans to minimize the worst-case expected operation cost during extreme events. The model utilizes a customized Benders decomposition algorithm. The results show that the proposed model can better balance economic efficiency and system security compared to traditional models.

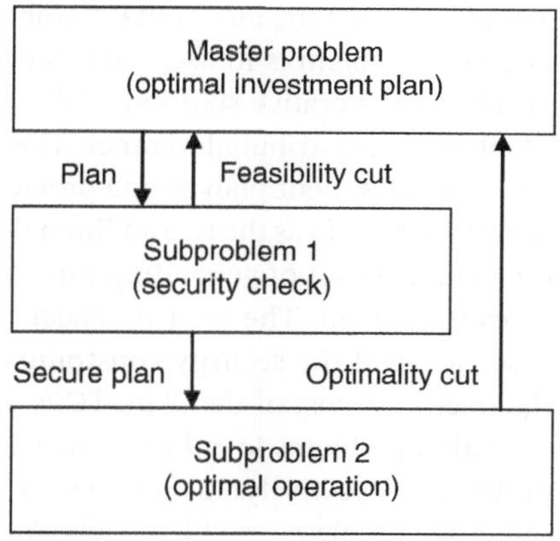

Figure 5.1 Proposed expansion planning using TC.

In Chapter 9, we presented a contingency planning model using TC. In this chapter, we present an expansion planning model using TC considering prevailing grid contingencies. Planning is a long-term problem that is solved over an extended period of time. TC is modeled as a short-term operation problem. The two problems are solved separately, considering the associated time periods. The critical point is to link these two problems so that the impact of TC on planning can be observed. This task is accomplished by using the security check and the optimal operation problems in Figure 5.1. The TC application could reduce the total planning cost and possibly defer the planning of the candidate units and transmission lines. In this situation, the transmission congestion in the system can vary. A corrective/preventive scheme is used to handle transmission lines and generation units contingencies. The security check problem performs corrective actions, while the preventive actions are considered in the Investment Problem. Our proposed approach uses a deterministic criterion to consider contingencies.

5.2 Economic Variables for Expansion Planning

Figure 5.1 depicts the proposed coordinated generation and transmission expansion planning model. The objective of this problem is to minimize the investment cost for new generating units and transmission lines, in addition to operating costs. This objective is subject to different security-based planning and operation constraints.

A Benders decomposition method is used to decompose the expansion planning problem. The optimality of Benders decomposition and its applicability to power system problems in practical cases are discussed in [34–37]. Also, in [38], the practical applications of Benders decomposition are discussed in detail, and it is explained that the most suitable approach to solving the multi-period expansion planning problem is to use Benders decomposition. The Benders decomposition provides the iterative solution of the mixed integer linear programming (MILP) problem in a distributed manner. The decomposed planning model includes the optimal investment plan as one problem and the security and the optimal operation of the system as the two additional problems. The Investment Problem utilizes the candidate set of generating units and transmission lines to find the optimal investment plan. The security check problem examines this plan to satisfy the base case and the security constraints. The optimal operation problem inspects the operating cost of the plan. TC is used in the security check and the optimal operation problems to enhance the feasibility and the optimality of the proposed problem, respectively. ILP is used to model the investment problem, while the other two problems are LP models. A typical set of planning constraints includes

- Capital investment funds
- Projected resource and line capacity
- Maximum number of generating units and transmission lines to be added
- Construction time of the candidate investments

A typical set of operation constraints includes

- Power balance
- Transmission line flow limits
- Generation limits

The approach is presented as follows:

5.2.1 Investment Problem

The Investment problem calculates the investment plan for generating units and transmission lines based on the initial set of candidates when considering generation and transmission expansion constraints. The lower bound for the original MIP planning problem is determined by the value of the problem objective at each iteration. The variables in this problem are binary, and the constraints are linear, so it is in the ILP format.

5.2.2 Security and Optimal Operation Problems

The security check problem checks whether the calculated investment plan satisfies the operation constraints. This problem would satisfy the power balance at every bus while preserving the base case and contingency constraints. If any of the constraints are violated, a feasibility cut is formed and added to the investment problem to solve the next iteration of the expansion planning problem. This iterative process will continue until a secure plan is achieved.

The optimal operation problem is used to check the minimum cost of the solution. This problem checks the optimality by calculating the upper bound of the original MIP planning problem's objective function and comparing it with its lower bound, which is already calculated in the investment problem. If the proposed plan is not optimal, Benders cuts will be formed and added to the investment problem to solve the next iteration.

5.2.3 Topology Control

TC is used in both security and optimal operation problems to minimize operating costs and transmission security violations. The TC binary variables are determined in the Investment problem and treated as constants in these two problems. Using TC states in the proposed cuts will help utilize candidate generating units and transmission lines in the planning options. The switchable line states appear as variables in the Investment problem, but they are governed by dual values in the security and optimal operation problems.

5.2.4 Cuts

In the proposed approach, two types of cuts are used. The first one is a feasibility cut, which is generated by the security check problem. The feasibility cut indicates that the security violations can be mitigated by readjusting the investment plan in each planning year as well as the state of switchable lines in the operation periods. This cut represents the coupling of the proposed investment plan and adjustments in the state of switchable lines. The dual variable in the feasibility cut is the incremental reduction in load balance violations. The second one is an optimality cut generated in the optimal operation problem. The optimality cut would limit the range of the investment problem objective to make it closer to the objective function of the expansion planning problem calculated prior to the decomposition. The optimality cut indicates that the objective value of the expansion planning problem can be decreased by modifying the investment plan in addition to the state of switchable lines. Similar to the feasibility cut, this cut represents the coupling of adjustments in the proposed investment plan and the state of switchable lines. The dual variables in the optimality cut are the incremental reduction in the objective function of the optimal operation problem.

5.2.5 Solution Procedure

The binary switching variables in the investment problem are governed by Benders cuts. On the other hand, the switching variables affect the investment plans. Therefore, the solution of the larger time scale is guided by the smaller time scale problem, and the solution of the smaller time scale problem is guided by the cuts produced at each iteration. The planning solution procedure is given as follows:

1) Solve the Investment problem by considering candidate generating units and transmission lines. There will be no Benders cut in the first iteration.
2) Given the proposed plan, check the feasibility of system constraints. If the security problem is feasible, proceed to Step 3. Otherwise, form the Benders cut and return to the investment problem.
3) Minimize the operating cost by considering the system constraints. Compare the solution value, i.e., an upper bound, with the objective of the investment problem, i.e., a lower bound. If the difference exceeds a predefined threshold, form the Benders cuts and return to the investment problem for the next iteration. Otherwise, consider the proposed plan as optimal.

5.3 An Economic Model for Resilience Planning in Infrastructure Expansion

The objective of the expansion planning problem is to minimize the total cost of the system while satisfying the system security and reliability constraints. The total cost includes the investment cost of new generating units and transmission lines plus the system operating cost. This objective is formulated as (5.1) and is subject to planning and operation constraints.

$$
\begin{aligned}
Min\ PP = {} & \sum_{t=1}^{NT}\sum_{i=1}^{CG}\frac{CI_{it}(X_{it}-X_{i(t-1)})}{(1+d)^{(t-1)}} \\
& + \sum_{t=1}^{NT}\sum_{j=1}^{CL}\frac{CI_{jt}(Y_{jt}-Y_{j(t-1)})}{(1+d)^{(t-1)}} \\
& + \sum_{t=1}^{NT}\sum_{b=1}^{NB}\sum_{i=1}^{NG}\frac{DT_{bt}OC_{ibt}PG_{ibt}}{(1+d)^{(t-1)}}
\end{aligned}
\tag{5.1}
$$

where t, i, j, and b are indices representing the year, generation unit, transmission line, and load block, respectively.

In the same order, NT, CG, CL, NB, and NG are the number of years, candidate generating units, candidate transmission lines, load blocks, and existing generation units. Additionally, within this framework, we use the following symbols: CI for the investment cost of a candidate generating unit or transmission line, X_{it} for the investment state of generating unit i in year t, d for the discount rate, Y_{jt} for the investment state of transmission line j in year t, DT_{bt} for the duration of load block b in year t, OC_{ibt} for the operating cost of generating unit i during load block

b in year t, and PG_{ibt} for the dispatched capacity of generating unit i during load block b in year t.

The solution to this problem would determine the size, location, and time for adding new generating units and transmission lines in an economical way that ensures the system capability to meet the anticipated load growth in the future. Using the Benders decomposition, the problem is decomposed: the investment problem provides the optimal plan, followed by two problems to provide solutions for the security check and the optimal operation.

5.3.1 Optimal Plan

The objective of the investment problem is to minimize the investment cost of new generating units and transmission lines.

$$Min\, MP$$

$$MP \geq \sum_{t=1}^{NT}\sum_{i=1}^{CG} \frac{CI_{it}(X_{it} - X_{i(t-1)})}{(1+d)^{(t-1)}} + \sum_{t=1}^{NT}\sum_{j=1}^{CL} \frac{CI_{jt}(Y_{jt} - Y_{j(t-1)})}{(1+d)^{(t-1)}} \tag{5.2}$$

This objective is subject to the planning constraints (5.3)–(5.12). The generating unit constraints include capital investment funds in a planning year as GCI_t in (5.4), anticipated resource capacity in a planning year with GUC_t as maximum added generating capacity in year t in (5.5), the maximum number of generating units to be added at a planning year as GUN_t in (5.6), and the construction time of the candidate investment as GCT_i in (5.7). Similarly, the transmission line constraints are capital investment funds in a planning year as LCI_t in (5.9), maximum transmission line capacity in a planning year as LUC_t in (5.10), and maximum number of transmission lines to be added in a planning year as LUN_t in (5.11), and the construction time of the candidate investment as LCT_j in (5.12). Constraint (5.3) specifies that once a generating unit is in place, its investment state will be 1 in the following years. The same constraint exists for transmission lines, as presented by (5.8).

$$X_{i(t-1)} \leq X_{it} \quad (i = 1, \ldots, NCG)(t = 2, \ldots, NT) \tag{5.3}$$

$$\sum_{i=1}^{CG} CI_{it}(X_{it} - X_{i(t-1)}) \leq GCI_t \quad (t = 1, \ldots, NT) \tag{5.4}$$

$$\sum_{i=1}^{CG} Cap_i(X_{it} - X_{i(t-1)}) \leq GUC_t \quad (t = 1, \ldots, NT) \tag{5.5}$$

$$\sum_{i=1}^{CG} (X_{it} - X_{i(t-1)}) \leq GUN_t \quad (t = 1, \ldots, NT) \tag{5.6}$$

$$X_{it} = 0 \quad if\, t < GCT_i \quad (i = 1, \ldots, NCG)(t = 1, \ldots, NT) \tag{5.7}$$

$$Y_{j(t-1)} \leq Y_{jt} \quad (j = 1, \dots, NCL)(t = 2, \dots, NT) \tag{5.8}$$

$$\sum_{j=1}^{CL} CI_{jt}(Y_{jt} - Y_{j(t-1)}) \leq LCI_t \quad (t = 1, \dots, NT) \tag{5.9}$$

$$\sum_{j=1}^{CL} Cap_j(Y_{jt} - Y_{j(t-1)}) \leq LUC_t \quad (t = 1, \dots, NT) \tag{5.10}$$

$$\sum_{j=1}^{CL} (Y_{jt} - Y_{j(t-1)}) \leq LUN_t \quad (t = 1, \dots, NT) \tag{5.11}$$

$$Y_{jt} = 0 \quad if\, t < LCT_j \quad (j = 1, \dots, NCL)(t = 1, \dots, NT) \tag{5.12}$$

The investment problem will determine the optimal investment plan and the lower bound for the original MIP planning problem. The optimality cuts will help minimize investment and operation costs in this problem. The states of switchable transmission lines are also examined in this problem. Here, the Benders cuts provide the state of switchable lines to the investment problem.

5.3.2 Security Check

Given the proposed plan for the investment problem, as well as the state of switchable lines, the security check problem minimizes the potential system violations. This task is accomplished by introducing slack variables in the power balance equation at each bus. The objective is to minimize the sum of non-negative slack variables.

$$Min\; v_t = \sum_{b=1}^{NB} \sum_{h=1}^{NH} (SL_{hbt,1} + SL_{hbt,2}) \tag{5.13}$$

where

$SL_{hbt,1}, SL_{hbt,2}$ are non-negative slack variables for power mismatch at bus h at load block b in year t

NH is the number of buses.

$$\sum_{i \in I_h} PG_{ibt}^{E} + \sum_{i \in I_h} PG_{ibt}^{C} - PD_{hbt} - \sum_{j \in J_h} PL_{jbt}^{E} - \sum_{j \in J_h} PL_{jbt}^{C} + SL_{hbt,1} - SL_{hbt,2} = 0 \quad (h = 1, \dots, NH)(b = 1, \dots, NB)(t = 1, \dots, NT) \tag{5.14}$$

Equation (5.14) shows the power mismatch at bus h. In this equation, existing and candidate lines are shown with superscripts E and C, respectively. Set J_h includes the lines connected to bus h, labeled as "to bus" or "from bus" in the set. PD_{hbt} represents load demand at bus h at load block b in year t while PL_{jbt} is power flow on the transmission line j at load block b in year t.

This problem is subject to existing and candidate-generating unit and transmission line constraints. These constraints represent the capacity of existing and candidate generating units (5.18)–(5.19), where Z_{jbt} represents the switching state of the transmission line j at load block b in year t, NEL is the number of existing transmission lines and UX_{ibt} is contingency state of the generating unit i at load block b in year t. Network constraints are represented by (5.20)–(5.23) for existing transmission lines, (5.24)–(5.27) for candidate transmission lines, and (5.28) for phase angle of reference bus, where γ_j is the susceptance of the transmission line j, δ_{mbt} is the voltage angle of bus m at load block b in year t, and UY_{jbt} is the contingency state of the transmission line j at load block b in year t.

$$X_{it} = \hat{X}_{it} \quad (i = 1, \dots, NCG)(t = 1, \dots, NT) \tag{5.15}$$

$$Y_{jt} = \hat{Y}_{jt} \quad (j = 1, \dots, NCL)(t = 1, \dots, NT) \tag{5.16}$$

$$Z_{jbt} = \hat{Z}_{jbt} \quad (j = 1, \dots, NEL)(b = 1, \dots, NB)(t = 1, \dots, NT) \tag{5.17}$$

$$0 \le PG^E_{ibt} \le PG^{max,E}_i UX^E_{ibt} \quad (i = 1, \dots, NEG)(b = 1, \dots, NB)(t = 1, \dots, NT) \tag{5.18}$$

$$0 \le PG^C_{ibt} \le PG^{max,C}_i X_{it} UX^C_{ibt} \quad (i = 1, \dots, NCG)(b = 1, \dots, NB)(t = 1, \dots, NT) \tag{5.19}$$

$$PL^E_{jbt} - \gamma_j \left(\delta^E_{mbt} - \delta^E_{nbt}\right) - M^E_j(1 - Z_{jbt}) - M^E_j \left(1 - UY^E_{jbt}\right) \le 0 \quad (j = 1, \dots, NEL)(b = 1, \dots, NB)(t = 1, \dots, NT) \tag{5.20}$$

$$PL^E_{jbt} - \gamma_j \left(\delta^E_{mbt} - \delta^E_{nbt}\right) + M^E_j(1 - Z_{jbt}) + M^E_j \left(1 - UY^E_{jbt}\right) \ge 0 \quad (j = 1, \dots, NEL)(b = 1, \dots, NB)(t = 1, \dots, NT) \tag{5.21}$$

$$PL^E_{jbt} \le PL^{max,E}_j Z_{jbt} UY^E_{jbt} \quad (j = 1, \dots, NEL)(b = 1, \dots, NB)(t = 1, \dots, NT) \tag{5.22}$$

$$-PL^E_{jbt} \le PL^{max,E}_j Z_{jbt} UY^E_{jbt} \quad (j = 1, \dots, NEL)(b = 1, \dots, NB)(t = 1, \dots, NT) \tag{5.23}$$

$$PL^C_{jbt} - \gamma_j \left(\delta^C_{mbt} - \delta^C_{nbt}\right) - M^C_j(1 - Y_{jt}) - M^C_j \left(1 - UY^C_{jbt}\right) \le 0 \quad (j = 1, \dots, NCL)(b = 1, \dots, NB)(t = 1, \dots, NT) \tag{5.24}$$

$$PL^C_{jbt} - \gamma_j \left(\delta^C_{mbt} - \delta^C_{nbt}\right) + M^C_j(1 - Y_{jt}) + M^C_j \left(1 - UY^C_{jbt}\right) \ge 0 \quad (j = 1, \dots, NCL)(b = 1, \dots, NB)(t = 1, \dots, NT) \tag{5.25}$$

$$PL^C_{jbt} \le PL^{max,C}_j Y_{jt} UY^C_{jbt} \quad (j = 1, \dots, NCL)(b = 1, \dots, NB)(t = 1, \dots, NT) \tag{5.26}$$

$$-PL^C_{jbt} \le PL^{max,C}_j Y_{jt} UY^C_{jbt} \quad (j = 1, \dots, NCL)(b = 1, \dots, NB)(t = 1, \dots, NT) \tag{5.27}$$

$$\delta_{ref} = 0 \tag{5.28}$$

In the above formulation, the constraints on candidate generating units and transmission lines include the associated binary variable, already determined in the investment problem. The state of switchable transmission lines is included in the existing transmission flow constraints (5.20)–(5.23). So, whenever a transmission line is switched off, its associated switchable state is zero, and the transmission line will be completely removed from the network. A subset of existing transmission lines is usually considered switchable. Therefore, for the remaining and non-switchable lines, the switching state (binary variable) will equal 1 in the investment problem. The transmission line j is between buses m and n with a transmission flow from bus m to bus n. In constraints (5.20), (5.21), (5.24), and (5.25), a large constant value, i.e., M, is used to satisfy the relaxation of associated constraints when they are eliminated. This disjunctive parameter would be larger than a minimum value and should not be very large.

The security check problem is solved for the base case and contingencies. During the contingencies, transmission lines can operate at their emergency ratings. So, PL_j^{max} will be adjusted accordingly. The contingency state of the element that is on outage will be zero. In the base case, this parameter will be one. The security check problem performs corrective actions to mitigate the violations. Those contingencies that cannot be mitigated with corrective actions are dealt with by preventive actions in the next iteration of the investment problem.

If the proposed objective is zero, the problem will proceed to the optimal operation problem. Otherwise, the Benders cut (5.29) will be formed and added to the investment problem for the next iteration.

$$v_t + \left\{ \begin{array}{l} \sum_{i=1}^{CG} \lambda_{it}(X_{it} - \hat{X}_{it}) + \sum_{j=1}^{CL} \mu_{jt}(Y_{jt} - \hat{Y}_{jt}) \\ + \sum_{j=1}^{NS} \sum_{b=1}^{NB} \pi_{jbt}(Z_{jbt} - \hat{Z}_{jbt}) \end{array} \right\} \leq 0 \tag{5.29}$$

where λ, μ, and π are dual values of constraints (5.15), (5.16) and (5.17), respectively.

This cut points out that the stated violations could be mitigated by changing the investment plan in addition to the state of switchable lines. In fact, this cut calculates the capacity signals for the investment of new generating units and transmission lines in case the existing ones cannot satisfy the system feasibility. The Benders cut will be formed for each violated case and added to the investment problem. The iterative procedure continues until a secure plan that satisfies the system feasibility is achieved in the base case and contingencies.

5.3.3 Optimal Operation

After satisfying the feasibility of the plan due to system constraints, the optimal operation will check the optimality of the solution cost. The objective of the optimal operation problem is to minimize the operating cost for every year and load block as

$$Min\, w_{bt} = \sum_{i=1}^{NG} \frac{DT_{bt}OC_{ibt}PG_{ibt}}{(1+d)^{(t-1)}} \tag{5.30}$$

This objective is subject to (5.15)–(5.28). The problem is solved for the base case in which contingency parameters equal 1. The solution provides the upper bound of the objective function of expansion planning. This upper bound is used to check the optimality of the solution. So, the stopping criterion is defined based on this solution. If the proposed plan is not optimal, a Benders cut will be formed and added to the investment problem for the next iteration. The proposed Benders cut (5.31) would restrict the lower bound of the objective function in the investment problem. Here, λ, μ, and π are dual values of (5.15), (5.16), and (5.17), respectively.

The iterative process between the investment problem and security and optimal operation problems will continue until an optimal solution to the expansion planning problem is calculated. The problem feasibility is ensured via the security check problem. At the same time, the optimality is guaranteed by comparing the solutions of the investment problem and the optimal operation problem at each iteration. The solution to the investment problem is the lower bound of the optimal solution. The upper bound of the optimal solution is found by using the results of the optimal operation problem. The optimal solution is obtained when the lower and upper bounds are close enough. These bounds form an effective convergence criterion in (5.32), where ε represents a small positive and predefined threshold.

$$\begin{aligned} MP \geq & \sum_{t=1}^{NT}\sum_{b=1}^{NB} \widehat{w}_{bt} + \sum_{t=1}^{NT}\sum_{i=1}^{CG} \frac{CI_{it}(X_{it} - X_{i(t-1)})}{(1+d)^{(t-1)}} \\ & + \sum_{t=1}^{NT}\sum_{j=1}^{CL} \frac{CI_{jt}(Y_{jt} - Y_{j(t-1)})}{(1+d)^{(t-1)}} \\ & + \sum_{t=1}^{NT}\sum_{i=1}^{CG} \lambda_{it}(X_{it} - \widehat{X}_{it}) \\ & + \sum_{t=1}^{NT}\sum_{j=1}^{CL} \mu_{jt}(Y_{jt} - \widehat{Y}_{jt}) \\ & + \sum_{t=1}^{NT}\sum_{j=1}^{NS}\sum_{b=1}^{NB} \pi_{jbt}(Z_{jbt} - \widehat{Z}_{jbt}) \end{aligned} \tag{5.31}$$

$$\frac{PP - MP}{PP + MP} < \varepsilon \tag{5.32}$$

The planning stage is performed yearly, while the operation stage is carried out for load blocks. The contingencies would last for the entire load block in a year. However, there are no limitations on the length of load blocks, which can be chosen for any period of time, from hours to months. The choice will be a tradeoff between the accuracy and the simplicity in the execution of the proposed model. However, the choice of load blocks would play an important role in the proposed model.

5.4 Case Studies

Two case studies consisting of a six-bus system and the IEEE 118-bus system are analyzed. The proposed method was implemented on a 2.4-GHz personal computer using CPLEX.

5.4.1 Six-Bus System

The six-bus system is shown in Figure 5.2. A 10-year planning horizon is considered. The system data for generating units and transmission lines are given in Tables 5.1 and 5.2, respectively. In Table 5.3, the forecasted yearly peak load is listed. This load is distributed at 40%, 30%, and 30% among buses 3, 4, and 5, respectively. To simplify the calculations, four load blocks are considered annually. The duration and quantity of load blocks in the first year are given in Table 5.4. The load blocks in subsequent years will change proportionately to those in Year 1. A set of four candidate generating units and four candidate transmission lines are considered as planning options in Tables 5.1 and 5.2. The construction time for generating units is considered to be three years, while it is less than one year for the transmission lines. It is assumed that there are no annual limitations on capital investments or the number of generating units and transmission lines. The

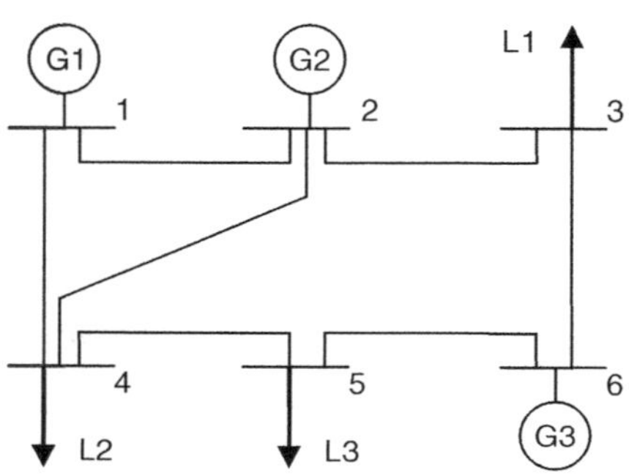

Figure 5.2 Six-bus system.

Table 5.1 Generating unit data of six-bus system.

Unit no.	Bus no.	Generating capacity (MW)	Investment cost ($/kW)	Operation cost ($/MWh)
1	1	100	Existing	15
2	2	100	Existing	18
3	6	50	Existing	23
4	1	100	200	15
5	2	80	270	21
6	2	60	250	24
7	3	20	250	24

Table 5.2 Transmission line data of six-bus system.

Line no.	From bus	To bus	*X* (p.u.)	Capacity (MW)	Investment cost ($/kW)
1	1	2	0.17	80	Existing
2	2	3	0.037	70	Existing
3	1	4	0.258	140	Existing
4	2	4	0.197	100	Existing
5	4	5	0.037	50	Existing
6	5	6	0.14	140	Existing
7	3	6	0.018	130	Existing
8	1	2	0.17	80	20
9	2	3	0.037	70	24
10	1	4	0.258	140	30
11	5	6	0.14	140	14

Table 5.3 Yearly peak load forecast of six-bus system.

Year	1	2	3	4	5
Peak (MW)	209	225	231	243	269
Year	6	7	8	9	10
Peak (MW)	297	307	315	324	330

Table 5.4 Load blocks in the first year.

Block	1	2	3	4
Duration (h)	87	2541	4380	1752
Load (MW)	209	192	167	150

discount rate is assumed to be 0. Three transmission lines, 1–4, 2–4, and 4–5 are considered switchable in the following four planning cases:

1) Case 1: Base case planning
2) Case 2: Transmission line 5–6 outage in load block 3 of year 4
3) Case 3: Generating unit 3 outage in load block 1 of year 6
4) Case 4: Simultaneous outages considered in Cases 2 and 3

The cases are discussed next.

Case 1: The 10-year expansion planning is considered without considering TC. According to this plan, line 2–3 is installed in the first year. Also, generating units 6 and 7 are added in years 5 and 8, respectively. In the first year, existing units 1, 2, and 3 try to satisfy the system load while maintaining the feasibility of transmission flows. Here, cheaper units 1 and 2 are dispatched at their maximum capacity, and the remaining load is supplied by unit 3. However, due to the congestion of line 2–3, unit 2 cannot increase its generation to its capacity. So, the more expensive unit 3 would increase its generation, which increases the operating cost. Accordingly, the candidate line 2–3 is installed at the first year. This installation increases the transmission capacity between buses 2 and 3, which allows unit 2 to increase its generation to its capacity in subsequent years.

The new system topology satisfies the system load while the system load is less than the total installed capacity, i.e., years 1–4. In year 5, the system would need to install a new generation capacity to help satisfy the additional load. So, unit 6 is placed at bus 2, which can transfer more power in conjunction with line 2–3 installation. Two options (i.e., either candidate unit 5 or 6) are available for installing a unit in bus 2. Unit 5 is more economical, while unit 6 requires a lower investment cost. Considering the remaining five years, the algorithm chooses unit 6. Similarly, in year 8, another unit is installed. The cheapest generating candidate from the operation viewpoint is unit 4, and the one from the planning viewpoint is unit 7. However, considering the remaining two years in the planning horizon, unit 7 is selected as being more economical. In this case, unit 5 is not a good option since the larger generation in bus 2 may lead to congestion on line 2–3. The total cost, in this case, is \$366.3M.

The TC application in Case 1 will result in a different expansion plan, where unit 4 is installed at year 5, and lines 2–3 and 1–4 are installed at years 1 and 9, respectively. Installing line 2–3 at year 1 will again relieve the congestion on line 2–3 and increase the dispatchability of units at buses 1 and 2. At year 5, unit 4 is chosen when the system needs to install additional generation capacity. This unit is the most economical candidate unit from an operation viewpoint. With the installation of unit 4 at bus 1, most of the system load will be supplied. The additional dispatch at bus 1 would lead to an increase in line flows with possible congestion. TC is used to mitigate the congestion. For example, the feasibility constraints cannot be satisfied when the load in year 5 is higher than 215 MW. The constraints are satisfied with the switching of line 2–4. Here, the total cost has dropped by 5.56% to $345.9M.

Case 2: In this case, the outage of line 5–6 in year 4 is considered. Similar to Case 1, line 2–3 is installed at year 1. With the possible outage of line 5–6, the installation of candidate line 5–6 will be necessary. Furthermore, candidate line 1–4 is installed in year 7 with the largest investment cost among candidate lines. However, the line will enhance the dispatch of energy generated by cheap units at bus 1. The total cost in this case is $347.9M, which is higher than that in Case 1.

Using TC, a similar expansion plan is proposed, i.e., the installation of unit 4 in year 5 and lines 2–3, 1–4, and 5–6 in years 1, 7, and 4, respectively. The line switching does not change the proposed plan in this case. However, the total cost is reduced to $346.4M. The lower total cost is mainly due to the switching of line 2–4. When the line is switched off, the flows on lines 1–2 and 1–4 will be less dependent on one another. So, both lines can transfer more power, resulting in the additional dispatch of units at bus 1. Since the cheapest units are at bus 1, the additional dispatch of these units will reduce the total cost.

Case 3: The outage of unit 3 at year 6 would change the proposed plan in Case 1. Like previous cases, new generation capacity is added at year 5. The proposed plan requires the installation of unit 7 at year 5 and, subsequently, unit 4 at year 6. The installation of unit 4 would increase the installed capacity. Also, unit 7 is added to satisfy the loads in buses 3 and 5. The installation of unit 7 is a preventive action to handle the outage of unit 3. The installed lines are 1–2, 2–3, and 1–4 in years 6, 1, and 6, respectively. Line 2–3 is installed to increase the dispatchability of units at buses 2 and 3, while lines 1–2 and 1–4 are installed to enhance the dispatchability of units at bus 1. The TC application would eliminate the previous installation of line 1–2 at year 6. The previous line flow violations that required the installation of line 1–2 are alleviated by the switching of lines 2–4 and 4–5. The total cost is $350.6M, which is 0.4% cheaper when using TC. This improvement is due to the elimination of candidate line 1–2 in the planning horizon and decreasing operating costs. This study demonstrates that applying TC could result in a similar but cheaper expansion planning.

Table 5.5 Candidate unit installation year of six-bus system.

Candidate unit	Case 1	Case 2	Case 3	Case 4
4	—	5	6	6
5	—	—	—	—
6	5	—	—	—
7	8	—	5	5

Table 5.6 Candidate unit installation year of six-bus system using TC.

Candidate unit	Case 1	Case 2	Case 3	Case 4
4	5	5	6	6
5	—	—	—	—
6	—	—	—	—
7	—	—	5	5

Table 5.7 Candidate line installation year of six-bus system.

Candidate line	Case 1	Case 2	Case 3	Case 4
1–2	—	—	6	—
2–3	1	1	1	1
1–4	—	7	6	6
5–6	—	4	—	4

Case 4: The expansion plan will not be influenced by TC when considering simultaneous outages of line 5–6 in year 4 and unit 3 in year 6. The expansion plan includes installing unit 4 at year 6, unit 7 at year 5, and lines 2–3, 1–4, and 5–6 at years 1, 6, and 4, respectively. The installation of unit 7 is a preventive action for the possible outage of unit 3, while the installation of line 11 is a preventive action for the possible outage of line 5–6. The TC application will not alter the proposed expansion plan but will reduce the total cost by 0.5%.

The simulation results for these four cases are summarized in Tables 5.5–5.7, and 5.8.

Table 5.8 Candidate line installation year of six-bus system using TC.

Candidate line	Case 1	Case 2	Case 3	Case 4
1–2	—	—	—	—
2–3	1	1	1	1
1–4	9	7	6	6
5–6	—	4	—	4

5.4.2 IEEE 118-Bus System

A modified IEEE 118-bus system is used to study the expansion planning problem with TC. The system has 118 buses, 54 units, and 186 branches. The data is given in motor.ece.iit.edu/data/Planning_118.xls. The candidate units and lines data are presented in Tables 5.9 and 5.10, respectively. The load growth rate is at 7%. The following three cases are analyzed:

Table 5.9 Candidate unit data of IEEE 118-bus system.

Unit no.	Bus no.	Generating capacity (MW)	Investment cost ($/kW)	Operation cost ($/MWh)
1	10	200	15	250
2	12	200	15	250
3	25	200	15	250
4	26	200	15	250
5	80	200	15	250
6	89	200	15	250
7	18	100	18	120
8	32	100	18	120
9	55	100	18	120
10	56	100	18	120
11	62	100	18	120
12	74	20	38	50
13	74	20	38	50
14	90	20	38	50
15	103	20	38	50
16	103	20	38	50

Table 5.10 Candidate transmission line data of IEEE 118-bus system.

Line no.	From bus	To bus	*X* (p.u.)	Capacity (MW)	Investment cost ($/kW)
1	30	38	0.054	100	30
2	77	82	0.0853	100	30
3	110	111	0.0755	100	30
4	20	21	0.0849	100	30
5	17	113	0.0301	100	30

Table 5.11 Switchable line data of IEEE 118-bus system.

Switchable line no.	Line no.	From bus	To bus	*X* (p.u.)	Capacity (MW)
1	30	23	24	0.0492	100
2	54	30	38	0.0540	100
3	65	47	49	0.0625	100
4	115	70	75	0.1410	100
5	151	80	97	0.0934	100
6	159	99	100	0.0813	100
7	164	100	104	0.2040	100
8	78	54	56	0.0096	100
9	90	60	61	0.0135	500
10	184	12	117	0.1400	100

1) Case 1: Base case planning
2) Case 2: Effect of discount rate on the planning solution
3) Case 3: Effect of the number of switchable lines on the planning solution

In Cases 1 and 3, the discount rate is assumed to be 10%, while in Case 2 the discount rate is changed to investigate its effect on the planning solution. In Cases 1 and 2, the effect of TC on planning is analyzed, where ten lines are considered switchable. The list of these switchable lines is presented in Table 5.11. In Case 3, the planning problem is solved for a variable number of switchable lines.

Case 1: The base case planning for the IEEE-118 bus system is solved as a coordinated generation and transmission expansion planning problem. At first, we disregard the TC option. The proposed plan for unit and line installations is shown

Table 5.12 Candidate unit installation year for IEEE 118-bus system in case 1.

Candidate unit	Without TC	With TC
1	—	7
2	5	5
3	7	6
4	5	—
5	6	5
6	—	—
7	—	—
8	10	—
9	9	10
10	10	10
11	—	10
12	10	10
13	10	10
14	10	10
15	—	—
16	—	—

Table 5.13 Candidate line installation year for IEEE 118-bus system in case 1.

Candidate line	Without TC	With TC
1	7	—
2	—	—
3	—	—
4	—	—
5	10	—

in Tables 5.12 and 5.13, respectively. In the first four years of planning, loads are satisfied, and there is no need to install any new units or lines. The installed generation capacity of the system is 5850 MW. In these four years, the congestion occurs in lines 8 and 96. However, in year 5, due to the congestion in line 38, the system cannot satisfy its load, even though it has enough installed capacity. To alleviate this congestion and transfer the required power to the loads, candidate units 2 and 4 are installed. The installations will mitigate the congestion of lines 8, 96, and 38, and the system will be able to meet the load. In the subsequent years 6 and 7, candidate units 5 and 3 are installed, respectively.

The installation of unit 5 in year 6 offers physical and economic advantages. On the other hand, installing unit 3 in year 7 will mitigate the congestion in lines 31, 38, and 51. The candidate units 2, 4, and 5 represent expensive investments necessary to provide sufficient generation capacity and feasibility of transmission flow. Also candidate line 1 is installed in year 7 to mitigate the congestion. Accordingly, new components will not be needed in year 8. However, in year 9, candidate unit 9 is installed, and subsequently, units 8, 10, 12, 13, and 14 are installed in the last planning year. Also, candidate line 5 is installed to increase the transfer capacity of the system. Considering the results, we conclude that the system is well-designed and the transmission lines offer a sufficient capacity to transfer the generated power most often.

TC provides quite different results. The candidate units 5 and 3 are installed one year earlier. The candidate units 4 and 8 are not installed, and instead, units 1 and 11 are installed at years 7 and 10, respectively. Also, the installation of unit 9 is delayed by one year, while the installation of other candidate units remains the same, i.e., unit 2 is installed in year 5, and units 10, 12, 13, and 14 are installed in year 10. Accordingly, the generation investment cost is decreased by 0.35%. TC eliminates the line investment in this case, as shown in Table 5.13. Also, TC enhances the power transfer capability by allowing existing transmission lines to transfer additional power. Here, lines 7, 8, 9, and 36 are congested in years 7–10. Also, line 96 is congested in years 1–5. The TC applications at certain load blocks will mitigate the congestion and enhance the system feasibility. Accordingly, the operating cost of the system is slightly increased. However, the proposed TC plan will reduce the total cost by 0.04%. As noted in the Introduction, TC may defer generation and transmission investments. Considering the obtained results, the installation of candidate unit 9 is deferred, while candidate units 3 and 5 are installed earlier than the base case solution.

Case 2: The sensitivity of the proposed plan to the discount rate is investigated. Generally, the discount rate is used to calculate the net present value and capacity payment for new generating unit and transmission line investments. It is clear that when the discount rate is higher, the investment candidates with higher capital costs become inferior. In the real world, investors may use different discount rates in their planning studies, depending on their financial situation and business strategies. However, in this case, the discount rate is assumed to be the same for all candidate units and lines. The discount rates of 10%, 5%, and 0% are considered. The candidate units and lines are the same as those of Case 1.

Table 5.14 Candidate unit installation year for IEEE 118-bus system in case 2.

Candidate unit	Discount rate (%)		
	10	5	0
1	7	—	—
2	5	4	4
3	6	—	—
4	—	5	4
5	5	6	4
6	—	—	—
7	—	10	4
8	—	10	4
9	10	9	4
10	10	9	4
11	10	10	4
12	10	10	4
13	10	10	4
14	10	10	4
15	—	—	—
16	—	—	—

The plan with a 10% discount rate was obtained in Case 1. By decreasing the discount rate to 5%, the candidate units with higher capital costs, i.e., units 1 to 6, are expected to become inferior. So, units 1 and 3 are not installed and installation of unit 5 is delayed by one year. However, Table 5.14 shows that cheaper units are installed in this case, in which units 7 and 8 are installed in year 10 and units 9 and 10 are planned in year 9. In addition, similar to the 10% discount solution, there would be no need to install new transmission lines when TC is applied, and the discount rate is 5%. Accordingly, the investment and total planning costs are increased by 14.4% and 0.63%, respectively.

If we further decrease the discount rate and set it to zero, a quite different investment plan will be obtained, as shown in the last column of Table 5.14. The discount rate of 0% means there is no difference in the installation year of the candidate unit or line since the investment cost will not change with time.

From the optimization viewpoint, all units are installed in the first year that they are required. Again using TS, there is no requirement for new transmission

Table 5.15 Candidate unit installation year for IEEE 118-bus system in case 3.

Candidate unit	Number of switchable lines				
	0	**10**	**20**	**30**	**40**
1	—	7	5	5	7
2	5	5	7	7	5
3	7	6	6	6	—
4	5	—	—	—	6
5	6	5	5	5	5
6	—	—	—	—	—
7	—	—	—	—	—
8	10	—	—	—	—
9	9	10	10	10	10
10	10	10	10	10	10
11	—	10	10	10	10
12	10	10	—	10	—
13	10	10	10	10	—
14	10	10	10	10	10
15	—	—	—	—	—
16	—	—	—	—	—

line investments, since TC satisfies the system feasibility. Compared with the 10% discount rate, the investment cost and the total planning cost are increased by 47.8% and 1.36%, respectively, when the discount rate is assumed to be zero.

Case 3: To examine the impact of TC on the planning results more comprehensively, the expansion planning problem in Case 1 is solved for a variety of switchable lines. The number of switchable lines varies from 0 to 40, where 0 means no line in the system has the switching capability and 40 means that more than one-fifth of the transmission system is considered switchable. A discount rate of 10% is considered for all cases.

The installation year of candidate units as a function of the number of switchable lines is shown in Table 5.15. The investment plan is a function of the number of switchable lines. However, no transmission line is installed when TC is applied. The investment cost is shown in Figure 5.3. This figure shows that the investment

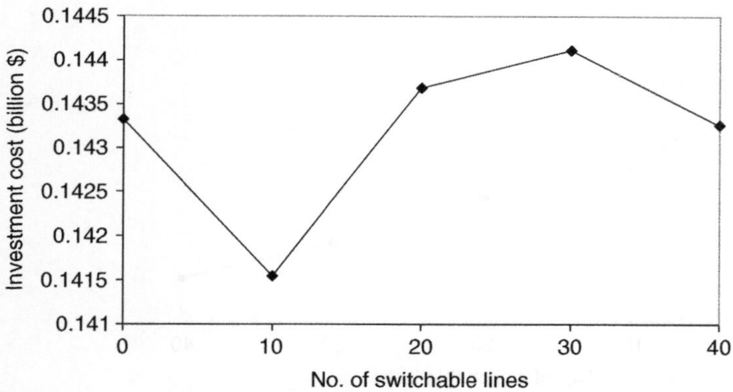

Figure 5.3 Investment cost in Case 3.

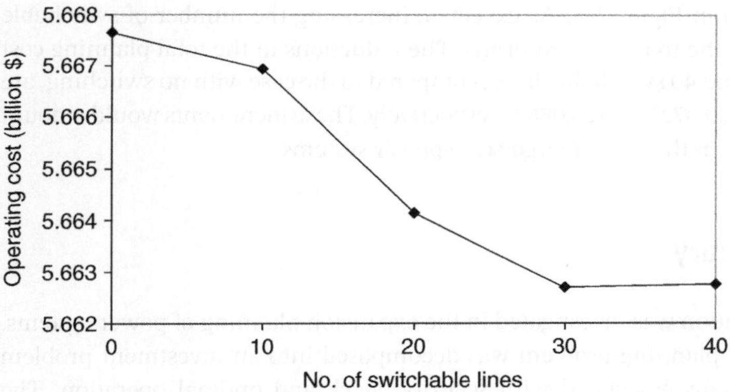

Figure 5.4 Operating cost in Case 3.

cost is a function of the number of switchable lines. In the case of the no-line switching capability, the investment cost includes the installation costs of units and lines; however, with the TC capability, the investment cost is only that of unit installations.

The operating cost of the system is shown in Figure 5.4. By increasing the number of switchable lines from 0 to 30, the operating cost is decreased. However, the operating cost of the 40 switchable lines is higher than that of 30.

So, by increasing the number of switchable lines, the operating cost of the system does not decrease monotonically.

The objective of the investment planning problem is to minimize the total planning cost, which is the sum of the investment and the operating costs. The total

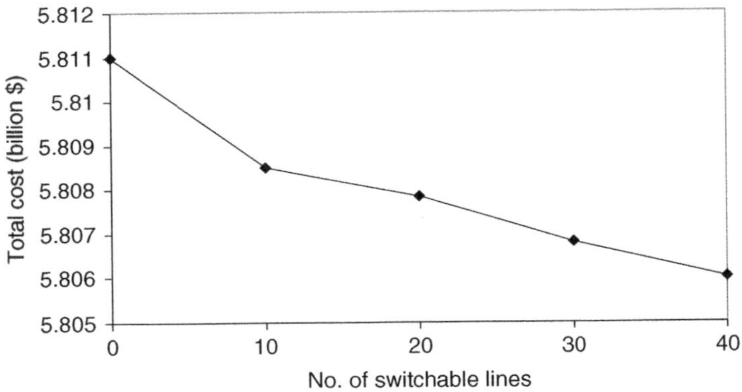

Figure 5.5 Total planning cost in Case 3.

cost is depicted in Figure 5.5. As expected, increasing the number of switchable lines decreases the total cost uniformly. The reductions in the total planning cost for 10, 20, 30, and 40 switchable lines, compared to the case with no switching, are 0.043%, 0.054%, 0.072%, and 0.086%, respectively. These increments would amount to large savings in the case of large-scale power systems.

5.5 Summary

The TC application was investigated in the expansion planning of power systems. The expansion planning problem was decomposed into an investment problem and two problems associated with security check and optimal operation. The investment problem found the optimal expansion plan by considering candidate units and lines. The security and optimal operation problems utilized in the proposed plan satisfy the feasibility and check the possibility of optimal operation. The TC application for enhancing the system security and economics was considered, while the role of TC for enhancing the expansion plan was considered in the investment problem. The proposed approach was analyzed further through numerical examples. The proposed TC approach can be utilized as an ISO model for coordinating the transmission expansion planning with the competitive generation capacity planning by representing an iterative process for representing the interactions among generation companies, transmission companies, and the ISO while considering the switching capability of lines.

References

1 L. L. Garver, "Transmission network estimation using linear programming," *IEEE Trans. Power App. Syst.*, vol. PAS-89, no. 7, pp.1688–1697, Sept./Oct. 1970.

2 S. T. Y. Lee, K. L. Hocks, and E. Hnyilicza, "Transmission expansion of branch-and-bound integer programming with optimal cost-capacity curves," *IEEE Trans. Power App. Syst.*, vol. PAS-93, pp. 1390–1400, Aug. 1974.

3 J. Choi, A. A. El-Keib and T. Tran, "A fuzzy branch and bound-based transmission system expansion planning for the highest satisfaction level of the decision maker," *IEEE Trans. Power Syst*, vol. 20, pp. 476–484, Feb. 2005.

4 J. Choi, T. Tran, A. A. El-Keib, R. Thomas, H. S. Oh, and R. Billinton, "A method for transmission expansion system expansion planning considering probabilistic reliability criteria," *IEEE Trans. Power Syst.*, vol. 20, pp. 1606–1615, Aug. 2005.

5 M. Lu, Z. Y. Dong, and T. K. Saha, "A framework for transmission planning in a competitive electricity market," *Transmission Distribution Conf. Exhib.*, Asia and Pacific, Aug. 2005, pp. 1–6.

6 G. B. Shrestha and P. A. J. Fonseka, "Congestion-driven transmission expansion in competitive power markets," *IEEE Trans. Power Syst.*, vol. 19, pp. 1658–1665, Aug. 2004.

7 O. B. Tor, A. N. Guven, and M. Shahidehpour, "Congestion-driven transmission planning considering the impact of generator expansion," *IEEE Trans. Power Syst.*, vol. 23, no.2, pp. 781–789, May 2008.

8 R. Romero and A. Monticelli, "A zero-one implicit enumeration method for optimizing investments in transmission expansion planning," *IEEE Trans. Power Syst.*, vol. 9, pp. 1385–1391, Aug. 1994.

9 S. Binato, M. V. Pereira, and S. Granville, "A new benders decomposition approach to solve power transmission network design problems," *IEEE Trans. Power Syst.*, vol. 16, pp. 235–240, May 2001.

10 P. Sánchez-Martín, A. Ramos, and J. F. Alonso, "Probabilistic Midterm Transmission Planning in a Liberalized Market," *IEEE Trans. Power Syst.*, vol. 20, no. 4, pp. 2135–2142, Nov. 2005.

11 E. L. Silva, H. A. Gil, and J. M. Areiza, "Transmission network expansion planning under an improved genetic algorithm," *IEEE Trans. Power Syst.*, vol. 15, pp. 1168–1175, Aug. 2000.

12 H. B. Sun and D. C. Yu, "A multiple-objective optimization model of transmission enhancement planning for independent transmission company (ITC)," in *Proc. IEEE Power Eng. Soc. Transmission Distribution Conf.*, Jul. 2000, vol. 4, pp. 2033–2038.

13 A. S. D. Braga and J. T. Saraiva, “A multiyear dynamic approach for transmission expansion planning and long-term marginal costs computation,” *IEEE Trans. Power Syst.*, vol. 20, pp. 1631–1639, Aug. 2005.

14 M. S. Kandil, S. M. El-Debeiky, and N. E. Hasanien, “A hybrid mathematical and rule-based system for transmission network planning in a deregulated environment,” in *Proc. IEEE Power Eng. Soc. Transmission Distribution Conf.*, Jul. 2001, pp. 1451–1456.

15 N. Yang and F. S. Wen, “A chance constrained programming approach to transmission system expansion planning,” *Electr. Power Syst. Res.*, vol. 75, pp. 171–177, Aug. 2005.

16 J. Contreras and F. F. Wu, "Coalition formation in transmission expansion planning,” *IEEE Trans. Power Syst.*, vol. 14, pp. 1144–1152, 1999.

17 T. De la Torre, J. W. Feltes, T. G. S. Roman and H. M. Merril, “Deregulation, privatization, and competition: Transmission planning under uncertainty,” *IEEE Trans. Power Syst.*, vol. 14, pp. 460–465, May 1999.

18 E. E. Sauma and S. S. Oren, “Economic criteria for planning transmission investment in restructured electricity markets,” *IEEE Trans. Power Syst.*, vol. 22, no.4, pp. 1394–1405, 2007.

19 M. Buygi, G. Balzer, H. Shanechi, and M. Shahidehpour, “Market-based transmission expansion planning”, *IEEE Trans. Power Syst.*, vol. 19, no. 4, pp. 2060–2067, Nov. 2004

20 M. O. Buygi, M. Shahidehpour, H. M. Shanechi, and G. Balzer, “Market based transmission expansion planning: Stakeholders’ desires,” in *Proc. 2004 IEEE Int. Conf. Electric Utility Deregulation, Restructuring Power Technol.*, 2004, vol. 2, pp. 433–438.

21 M. O. Buygi, M. Shahidehpour, H. M. Shanechi and G. Balzer, “Market based transmission planning under uncertainties,” *Int. Conf. Probabilistic Met. Appl. Power Syst.*, Sep. 2004, pp. 563–568.

22 M. O. Buygi, M. Shahidehpour, H. M. Shanechi, and G. Balzer, “Market based transmission expansion planning: Fuzzy risk assessment,” in *Proc. 2004 IEEE Int. Conf. Electric Utility Deregulation, Restructuring Power Technol.*, 2004, vol. 2, pp. 427–432.

23 J. Buechler, “Transmission planning in a market based environment,” in *Proc. 2005 IEEE Power Eng. Soc. General Meeting*, 2005, pp. 1850–1852.

24 J. H. Roh, M. Shahidehpour, and Y. Fu, “Security-constrained resource planning in electricity markets,” *IEEE Trans. Power Syst.*, vol. 22, no. 2, pp. 812–820, May 2007.

25 J. H. Roh, M. Shahidehpour, and Y. Fu, “Market-based coordination of transmission and generation capacity planning,” *IEEE Trans. Power Syst.*, vol. 22, no. 4, pp. 1406–1419, Nov. 2007.

26 E. Sauma and S. Oren, "Proactive planning and valuation of transmission investments in restructured electricity markets," *J. Regul. Econ.*, vol. 30, pp. 261–290, Sep. 2006.

27 S. de la Torre, A. J. Conejo, and J. Contreras, "Transmission expansion planning in electricity markets," *IEEE Trans. Power Syst.*, vol. 23, no. 1, pp. 238–248, Feb. 2008.

28 L. P. Garces, A. J. Conejo, R. Garcia-Bertrand, and R. Romero, "A bilevel approach to transmission expansion planning within a market environment," *IEEE Trans. Power Syst.*, vol. 24, no. 3, pp. 1513–1522, Aug. 2009.

29 Y. Wang, S. Liu, J. Wang, and B. Zeng, "Capacity expansion of wind power in a market environment with topology control," in *IEEE Trans. Sustain. Energy*, vol. 10, no. 4, pp. 1834–1843, Oct. 2019, doi: https://doi.org/10.1109/TSTE.2018.2873646.

30 X. Li and Q. Xia, "Transmission expansion planning with seasonal network optimization," in *2020 IEEE Power Energy Soc. Innovative Smart Grid Technol. Conf. (ISGT)*, Washington, DC, USA, 2020, pp. 1–5, doi: https://doi.org/10.1109/ISGT45199.2020.9087743.

31 Y. Hu, M. Wang, X. Lian, S. Liu, and Q. Zhang, "A reduced disjunctive model for transmission expansion planning considering unit commitment and optimal transmission switching," in *2021 IEEE Int. Conf. Electr. Eng. Mechatr. Technol. (ICEEMT)*, Qingdao, China, 2021, pp. 623–627, doi: https://doi.org/10.1109/ICEEMT52412.2021.9601896.

32 X. Cao, J. Wang, J. Wang, and B. Zeng, "A risk-averse conic model for networked microgrids planning with reconfiguration and reorganizations," *IEEE Trans. Smart Grid*, vol. 11, no. 1, pp. 696–709, Jan. 2020, doi: https://doi.org/10.1109/TSG.2019.2927833.

33 Y. Yuan, H. Zhang, H. Cheng, J. Zhang, and Z. Wang, "Transmission and energy storage expansion planning against extreme event," *2022 IEEE 5th Int. Electr. Energy Conf. (CIEEC)*, Nangjing, China, 2022, pp. 4347–4352, doi: https://doi.org/10.1109/CIEEC54735.2022.9845866.

34 M. Shahidehpour and Y. Fu, "Benders decomposition," *IEEE Power Energy Mag.*, vol. 3, no. 2, pp. 20–21, Mar. 2005.

35 M. Shahidehpour and V. Ramesh, "Nonlinear programming algorithms and decomposition strategies for OPF," in *IEEE/PES Tutorial Optimal Power Flow*. Piscataway, NJ: IEEE Press, 1996.

36 H. Ma and M. Shahidehpour, "Transmission constrained unit commitment based on benders decomposition," *Electr. Power Energy Syst.*, vol. 20, no. 4, pp. 287–294, Apr. 1998.

37 A. M. Geoffrion, "Generalized benders decomposition," *J. Optim. Theory Appl.*, vol. 10, no. 4, pp. 237–261, 1972.

38 Granville S, Pereira MVF, "Analysis of the linearized power flow model in Benders decomposition," Stanford University, Techn. Rep. SOL 85-04, 1985.

6

The Economics of Operational Risk Management in the Face of Natural Disasters

6.1 Principles of Operations Management Against Wildfires

The keystone of modern society is a resilient power grid that ensures an uninterrupted electricity supply to citizens and interdependent lifeline systems even during extreme events. Legacy power grids, however, have failed their customers following many recent disasters while largely contributing to some of these events. The 2018 Camp fire in Northern California—which was sparked by Pacific Gas and Electric Company (PG&E)'s electrical infrastructure—took 85 lives, burned a total of 153,336 acres, destroyed more than 18,800 homes and structures, and made PG&E face a multibillion-dollar lawsuit as a result [1, 2]. The 2021 Dixie Fire in Northern California is a microcosm of recent catastrophic wildfires caused by power lines, which burned over 963,000 acres and destroyed thousands of structures [3]. This devastating fire not only caused extensive damage to the power infrastructure but also led to widespread outages, as utility companies were forced to shut down parts of the grid and implement pre-emptive measures, such as Public Safety Power Shutoffs (PSPS), to prevent further spread of the fires [4]. To better understand, Figure 6.1 highlights the growing trend of wildfires ignited by power grids over time, among other causes of wildfires in the United States from 1992 to 2020—emphasizing the increasing importance of mitigating these incidents. The chart in Figure 6.2 compares the frequency to the average burned area (impact) for different causes of wildfire incidents in the United States from 1992 to 2020. As shown, the impact of wildfires caused by the grid is significant [5].

While power grids are responsible for less than 10% of wildfire ignitions, they are responsible for more than 90% of fatal wildfires in Australia. In California, power lines are responsible for igniting four out of 20 most massive wildfires by acreage burned and five out of 20 largest wildfires by structures destroyed. The scale of these wildfires highlights their significance, as their impact on other parts of the grid expands beyond their initial ignition point, correlating with the

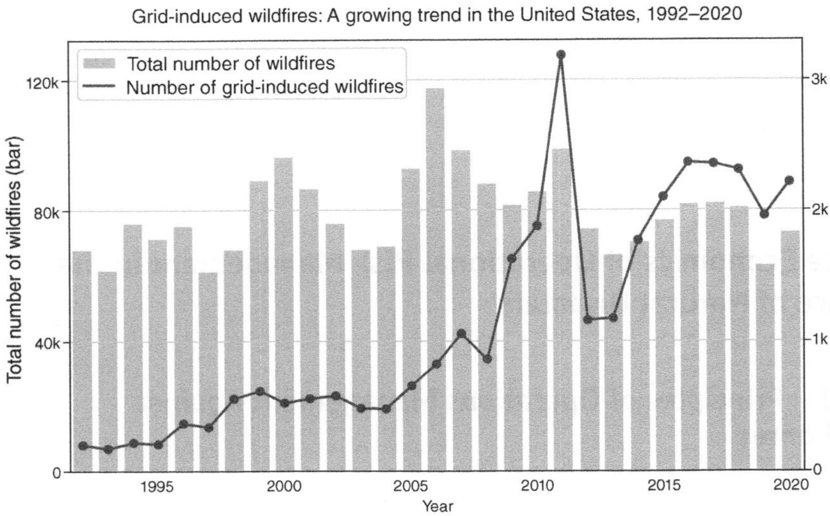

Figure 6.1 Trend of wildfires caused by the grid over time. Source: Ref. [5]/U.S. Department of Agriculture/Public domain.

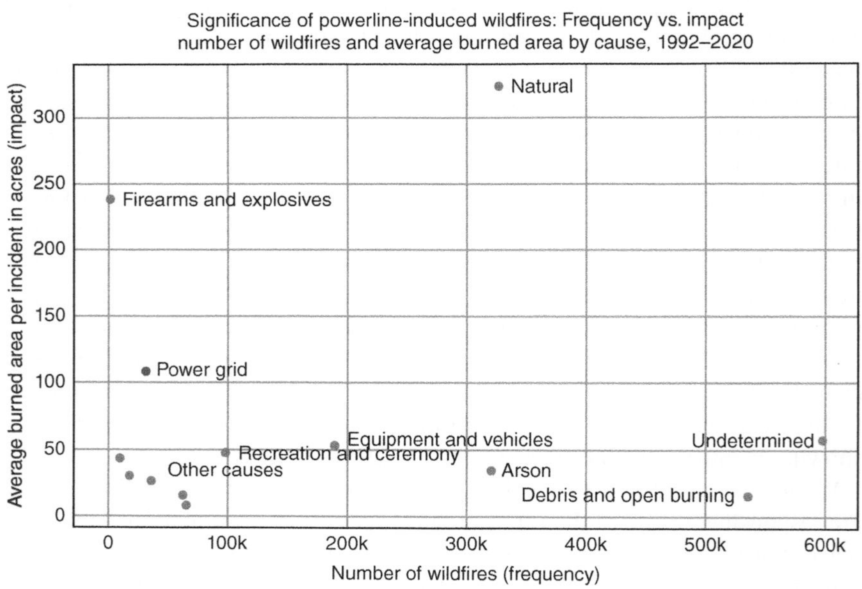

Figure 6.2 Frequency and impact comparison for wildfire causes. Source: Ref. [5]/U.S. Department of Agriculture/Public domain.

extent of exposed infrastructure. Therefore, understanding the risk, vulnerability, and impact of these extreme events on transmission lines and nodes is crucial for power system planners to develop effective strategies for managing grid resilience.

There has been a significant increase in research at the intersection of power grids and wildfire risk management. Several studies have addressed grid-ignited wildfire prevention and risk mitigation strategies. Studies in [6–10] investigate measures adopted by power utilities for wildfire prevention and response. These measures encompass monitoring and clearing vegetation near transmission lines, fault detection strategies, equipment maintenance, enhanced inspection and monitoring standards, various grid hardening actions, and applying PSPS and operational best practices. Authors in [11] present a framework using historical wildfire data to quantify power line risks. This framework employs an optimization model based on the determined risk map to prioritize which overhead lines should convert to underground cables, reducing network wildfire risk. Taking a more data-driven approach, [12] combines the latest electrical protection technology, statistical forecasting, meteorological data, and regional risk modeling to promote a proactive wildfire prevention strategy, guiding utilities in improving grid operations. The study in [13] evaluates PSPS as a countermeasure against grid-induced wildfires during adverse weather conditions, providing a methodology for selecting specific network segments for de-energization to reduce risk. Expanding upon this theme, the study in [14] offers an optimization model based on wildfire risks and efficient post-event restoration in regions vulnerable to wildfires to guide utilities on targeted power shutoffs, updating recommendations based on fresh forecast data. Authors in [15] present an optimization model designed to reduce wildfire risks by enhancing grid operations during PSPS events and determining the best options for placing and investing in infrastructure, including grid-scale batteries, solar PV, and line-hardening measures. From a cost-benefit standpoint, [16] quantifies potential monetary losses due to wildfires and contrasts them with the costs of fire-hardening retrofits. It evaluates the economic implications of retrofitting transmission lines in San Diego County to mitigate wildfire risks.

Transitioning from wildfire prevention and mitigation strategies, other studies have focused on developing methods for early detection of grid-induced wildfires. The rising severity of wildfires, as pointed out in [17], is partially linked to faults in electric utility distribution circuits. Building on this observation, the study utilizes waveform analytic techniques and real-time monitoring to introduce a real-time diagnostic technique designed to detect potential fire ignition mechanisms before they escalate. [18] delves deeper into understanding vegetation ignition characteristics and, using real-world data, proposes a Hybrid Step XGBoost (HSXG) model to pinpoint hazardous powerline-vegetation contacts, emphasizing early wildfire detection. Meanwhile, in [19], a shift is seen toward harnessing the power of real-time deep learning algorithms to accurately classify and locate faults in power distribution networks that may trigger potential fires. Similarly, [20] targets early detection, presenting a methodology geared toward precisely predicting wildfire ignition risks, with a special focus on high impedance faults.

Another critical avenue of research focuses on grid resilience and adapting operations under the stress of wildfire incidents. The study in [21] explains that the challenges of enhancing resilience against wildfires include real-time responses, situational awareness, preventive measures, proactive corrections, and dynamic restoration strategies. Building on this, [22] presents a proactive generation redispatch strategy to enhance power grid resilience during wildfires, utilizing a Markov decision process (MDP). In [23], the authors present a framework to assess the impact of electric vehicle (EV) evacuations and associated charging demands on grid resilience amid wildfire events, underscoring the growing significance of power system planning and operations during such extreme events. The study in [24] proposes an optimization-based approach for grid expansion planning to balance both wildfire risk and reliable power provision, focusing on adding new power lines, modifying existing lines, and integrating DERs. Other studies, including [25, 26], as well as [7] and [9], have focused on the role of DERs, microgrids, grid-hardening, and grid modernization to enhance grid resilience and minimize the potential impacts of wildfire on the systems. The study in [27] examines the physical byproducts of wildfires, such as heat, smoke, and ash, and formulates models to assess their impacts on power lines, renewable energy sources, and vital electrical infrastructure that can aid in the management and adaptation of the grid during wildfires. Similarly, [28] investigates how the intense heat from wildfires affects the operational capacity of power lines in their proximity, leading to potential disruptions and infrastructure degradation.

Shifting the focus to grid-induced wildfire risk assessment, a pivotal aspect of this chapter, the literature offers different techniques and methodologies. Using historical data as a foundation, [11] employs wildfire risk maps and powerline locations to assess the risk of powerline-induced wildfire ignition. Building upon the historical approach, [12] integrates machine learning and past wildfire incidents to formulate a time series forecasting model for estimating the risk of wildfires in relation to power systems. Similarly, [29] leverages machine learning models combined with weather, vegetation, and terrain data to pinpoint wildfire risk probabilities in powerline-adjacent regions. Technical factors play a role in risk prediction as well. For instance, [30] highlights a notable correlation between escalating wind speeds and powerline system failures, which can instigate wildfires. Delving deeper into predictive measures, [13] adopts a surrogate machine learning model to anticipate wildfire ignition risks for individual powerlines under extreme climatic conditions. Broadening the scope, [31] integrates wildfire simulation with forest landscape composition modeling to assess wildfire threats to a powerline corridor. Finally, [32] merges historical insights with real-time monitoring to devise a comprehensive method for evaluating wildfire dangers within power transmission line corridors.

6.2 Variables for Quantifying Risk and Vulnerability Against Wildfires

The methodology for developing the framework in this study consists of several steps that integrate a wildfire spread simulator with power flow analysis. The primary goal is to quantify risk and vulnerability, identify areas prone to wildfires, and determine high-risk components. The proposed steps are as follows:

6.2.1 Closing the Data Gap

The first step in the methodology involves the acquisition of essential data sets, which are crucial for both the execution of the Fire Area Simulator (FARSITE) simulator and the subsequent analysis of its impact on the power network. These datasets include network topology, weather conditions, and landscape fuel data for the region under consideration (Figure 6.3).

The first layer of the dataset consists of network topology data, which maps the structure of the power grid, detailing the locations and connections of nodes and lines. This specific information is crucial not only for identifying which lines might be tripped and which nodes could be impacted by the spread of wildfires but also for analyzing how power flow is dispatched across the network and how it may be affected by wildfire spread in post-simulation analysis.

The second layer encompasses weather data, which, without loss of generality, could be historical or synthetically generated using statistical parameters for extreme conditions. This feature data is a statistically significant predictor of the ignition and spread of wildfires and includes factors such as temperature, humidity, wind speed, and wind direction of the investigated region, which are all integral to accurately simulating wildfire spread and behavior [33, 34]. Specifically, these parameters are essential inputs for the FARSITE wildfire simulator, ensuring it captures the dynamic nature of such incidents comprehensively.

The third layer comprises landscape and fuel data, essential for capturing a realistic representation of wildfire behavior, which includes topographic details of the landscape such as elevation, slope, solar aspect, and barriers, along with landscape fuel properties including fuel flammability model, canopy cover percentage,

Figure 6.3 The multi-layered data foundation for the framework.

crown height, crown base height, and crown bulk density. Each parameter plays a unique role in the behavior and progression of wildfires [34]. For instance, elevation and slope directly affect the speed and direction of fire spread, with fires generally moving faster uphill due to the preheating of fuels positioned above the flames. The solar aspect determines which part of the landscape is most exposed to sunlight, influencing fuel dryness and, therefore, its combustibility. Barriers can interrupt or redirect the movement of fires. Properties such as the fuel flammability model and canopy cover percentage influence how easily a fire ignites and spreads through specific vegetation types. The crown height, base height, and bulk density are crucial for predicting the potential of crown fires, which rapidly propagate through the canopy or treetops and cause spotting. This can allow the fire to jump barriers and ignite new fires away from the main blaze. As a comprehensive simulator, FARSITE requires these detailed inputs to produce accurate simulations that capture the complex behavior of wildfires. Ensuring this data accurately represents the region is crucial for generating reliable wildfire spread simulations. Collecting these layers provides a solid foundation for the subsequent steps of the methodology.

6.2.2 Wildfire Scenarios Generation

The second step involves generating diverse wildfire ignition scenarios, simulating the conditions under which hypothetical wildfires originate and propagate. These scenarios should be designed to cover a broad range of possibilities for the network under study to thoroughly analyze the risks and vulnerabilities of its components. As previously highlighted, weather conditions, such as wind speed and direction, temperature, and humidity, can drastically alter wildfire behavior, and their variations must be incorporated into these scenarios. Furthermore, ignition locations must be strategically placed on transmission lines and high-risk areas of the network to simulate grid-caused wildfire ignition. These points can be positioned at equal intervals on the lines for systematic coverage. Alternatively, by leveraging risk maps, additional points can be placed in identified network hotspots, emphasizing areas with a higher likelihood of fire origination. Together, these parameters form the basis of the wildfire scenarios and contribute to a comprehensive representation of potential wildfire events. The extensive suite of scenarios allows for an in-depth exploration of how wildfires could interact with and impact the power grid.

6.2.3 Wildfire Spread Simulation

In this step, wildfire scenarios and landscape fuel data are fed into the FARSITE simulator to determine the areas affected by wildfire in each scenario. Developed by Mark Finney, FARSITE models the spread of wildfires using a detailed spatial-temporal and deterministic simulation process. The modeling approach of FARSITE utilizes Huygens' principle of wave propagation to simulate the advancement of a fire front. The model incorporates several sub-models of fire behavior, including surface fire, crown fire spread, fire acceleration, fuel moisture, and spotting from torching trees. All these calculations pertain to the fire's perimeter, primarily advancing the fire edge according to Huygens' principle. FARSITE dynamically divides the simulation into discrete time steps. Within each step, the model calculates the progression and behavior of wildfire, represented as a continuously expanding polygon defined by a series of vertices. FARSITE structures its simulation through a series of timesteps, wherein each timestep computes the growth of a fire polygon by aggregating spread from its vertices. This holistic growth is shaped by the fire's Rate of Spread (ROS) at each vertex of the fire edge. Instead of a fixed timestep, FARSITE dynamically modifies it based on the distance resolution, which signifies the furthest horizontal spread allowed before new landscape data is required. To maintain the integrity of the fire's shape, new ones are inserted if two vertices on the fire perimeter get too distant. The aim is to ensure the fire's depiction remains true to the model's parameters while accommodating varying environmental conditions. This continuous adjustment optimizes the simulation's accuracy, ensuring the fire's depiction is both detailed and consistent across diverse terrains. FARSITE's simulation process also corrects overlapping fire growth, calculates fire area and perimeter, and merges separate fires as needed. However, like any deterministic simulator, FARSITE can encounter inaccuracies, mainly due to challenges in securing reliable input data with the required spatial and temporal resolution. Despite this, the simulator has proven to be a robust tool, particularly with accurate data. Numerous federal land management agencies recognize FARSITE as an essential instrument for predicting fire growth [35]. The choice of FARSITE for our framework is driven by its proven reliability and capability to model intricate wildfire interactions in diverse terrains. The detailed output from FARSITE simulations depicts the intensity, progression, and extent of each wildfire scenario, identifying areas of the power grid directly impacted by the wildfires. This provides a detailed dataset for subsequent steps in the methodology.

6.2.4 Impact Analysis

Following the wildfire spread simulation, a data-driven impact analysis is conducted to evaluate the repercussions of each scenario on the grid. The output from the FARSITE simulator, which includes detailed wildfire perimeters, fire intensity, and fire arrival time, is integrated with the spatial coordinates of grid components. Through spatial analysis, the lines and nodes exposed to the wildfire in each simulation scenario are identified. These affected components represent potential points of failure or disruption in the grid. The impact analysis also considers the severity and duration of these impacts based on the output of wildfire simulation at each location. The result of this step is a comprehensive mapping of which grid components are likely to be impacted in a given scenario, serving as an input for subsequent steps in the methodology. Figure 6.4 illustrates the spatial intersection of the simulated wildfire spread output overlaid by the geographical layout of the grid, showing the components directly affected by the fire.

6.2.5 Network Modification

6.2.5.1 Impacted Components Removal

Once the impacted lines and nodes are identified, the next step involves modifying the grid's component specifications based on the impact analysis results for each scenario. This modification is intended to simulate the changes in the operational

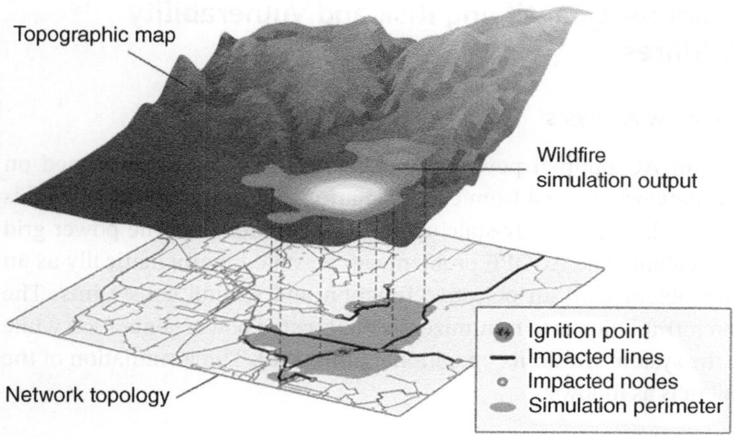

Figure 6.4 Wildfire simulation output mapped with the network topology.

state of the grid due to the faults caused by wildfires. For instance, a transmission line intersecting with a wildfire spread may be considered damaged or at risk of failure and thus may be rendered non-operational in the modified network. Similarly, nodes within the fire perimeter might experience reduced capacity or total outage depending on the extent of the fire impact. These modifications are then integrated into the power grid model, resulting in a modified network. This approach ensures that the subsequent power flow analysis considers the potential impacts of wildfire in each scenario on the network's operations.

6.2.5.2 Virtual High-Cost Generators Placement

In the power flow analysis, virtual high-cost generators are placed at each node to simulate potential outages (i.e., load curtailment). These generators are considered "virtual" because they do not represent actual power sources within the network but rather function as placeholders for situations where a real source isn't available due to a wildfire-induced outage. These generators are designated as "high-cost" to reflect the undesirable nature of power outages. Under normal operating conditions, the power system is always inclined to utilize lower-cost actual power generators. Resorting to these high-cost virtual generators in the model signifies that the system is under stress and must use more expensive means, here load curtailment, to maintain a stable power flow. Therefore, they act as a useful tool in the simulations to understand the magnitude of power outages at each node under various wildfire scenarios and inform the placement of real power sources, such as microgrids and DERs, to increase resilience.

6.3 A Model for Quantifying Risk and Vulnerability Against Wildfires

6.3.1 Power Flow Analysis

In this phase, an AC optimal power flow (ACOPF) analysis is performed on each modified network derived from each scenario. The primary purpose of this analysis is to calculate the steady-state operational conditions of the power grid post-wildfire incident. The ACOPF problem can be stated mathematically as an optimization problem with an objective function and several constraints. The most common formulation is to minimize the total cost of power generation while maintaining the system within its operational limits [36]. The formulation of the ACOPF problem is as follows:

$$min \sum_{i}^{G} C_i\left(P_i^g\right) \tag{6.1}$$

where

$C_i\left(P_i^g\right)$ represents the cost of power generation by generator i
G is the set of all generators

Our variables for optimization are the real and reactive power outputs from each generator (P_i^g and Q_i^g) and the voltages at each bus (V_n). The main constraints we need to satisfy are:

Power Balance Constraints: This ensures that the total generated power plus the power injected through transmission lines equals the total power demanded at each node, separately for both real and reactive power. In mathematical terms:

$$P_n^g - P_n^d + P_n^l = 0 \quad \forall n \in N \tag{6.2}$$

$$Q_n^g - Q_n^d + Q_n^l = 0 \quad \forall n \in N \tag{6.3}$$

where

P_n^g and Q_n^g are the real and reactive power generated at node n
P_n^d and Q_n^d are the real and reactive power demand at node n
P_n^l and Q_n^l are the real and reactive power flows injected to node n
N is the set of all nodes.

Generation Constraints: Each generator's output should not exceed its maximum capacity nor fall below its minimum.

$$P_i^{min} \leq P_i^g \leq P_i^{max} \quad \forall i \in G \tag{6.4}$$

$$P_i^{min} \leq P_i^g \leq P_i^{max} \quad \forall i \in G \tag{6.5}$$

where P_i^{min}, P_i^{max}, Q_i^{min}, and Q_i^{max} are respectively the minimum and maximum limits for real and reactive power for generator i.

Voltage Constraints: The voltage magnitude at each bus should stay within a specified range to ensure safe and efficient operation.

$$V_n^{min} \leq |V_n| \leq V_n^{max} \quad \forall n \in N \tag{6.6}$$

where

V_n is the voltage at node n

V_n^{min} and V_n^{max} are respectively the minimum and maximum permissible voltage magnitudes at node n.

Power Flow Constraints: The power flow in each transmission line should not surpass its capacity to prevent line overheating and potential failures.

$$-S_j^{min} \leq S_j \leq S_j^{max} \quad \forall j \in L \tag{6.7}$$

where

S_j is the complex power flow of line j

S_j^{min} and S_j^{max} are respectively the minimum and maximum permissible power flow of line j

L is the set of all transmission lines.

In this process, using virtual high-cost generators at each node guarantees the feasibility of power flow, even under the stress of component outages. Thus, the power flow analysis provides insights into how the grid should optimally adapt and respond to the impacts of wildfire incidents. The outcome of the ACOPF analysis offers vital information on power flows, voltages, generator outputs, and outages for each scenario. This data reveals the power grid's performance under different wildfire conditions, helping quantify measures of risk and vulnerability.

6.3.2 Data Analysis

Once the power flow analysis is completed for all scenarios, the next step is to analyze the data obtained to derive meaningful metrics.

6.3.2.1 Vulnerability

Vulnerability is defined as the potential susceptibility of the system components to the effects of grid-induced wildfire. The vulnerability of a particular node (U_n^{node}) or line (U_j^{line}) can be quantified by counting the times it is affected across all scenarios.

$$U_n^{node} = \frac{1}{|Z|} \sum_{s \in Z} A_{ns} \tag{6.8}$$

$$U_j^{line} = \frac{1}{\left|\overline{Z}_j'\right|} \sum_{s \in Z_j'} A_{js} \tag{6.9}$$

where A_{ns} and A_{js} are binary values representing whether node n and line j are affected in scenario (n) out of all scenarios (Z) or scenarios not induced by line $j\left(\overline{Z}_j'\right)$.

A line is considered affected when it falls within the wildfire simulation's perimeter, and a node is considered affected when a line connected to it or the node itself is caught in the perimeter.

6.3.2.2 Risk

Risk is defined as the potential impact on the overall grid due to the failure of a particular component (in this case, a transmission line). The risk associated with each line (R_j) can be calculated by taking the average of power outages in a subset of wildfire scenarios $\left(Z_j'\right)$ which are induced by line j. In this equation, P_{jsn}^o represents the amount of power outage caused by line j in scenario s at node n.

$$R_j = \frac{1}{\left|Z_j'\right| |N|} \sum_{s \in Z_j'} \sum_{n \in N} \frac{P_{jsn}^o}{P_n^d} \tag{6.10}$$

6.3.2.3 Impact

Similar to Eq. (6.10), the impact metric associated with each node (I_n) can be defined as the average power outage at each node in the system, considering all scenarios.

$$I_n = \frac{1}{|Z||L|} \sum_{s \in Z} \sum_{j \in L} \frac{P_{jsn}^o}{P_n^d} \tag{6.11}$$

In essence, Eq. (6.11) offers a holistic view of how power outages can impact a node across an ensemble of wildfire scenarios.

6.4 A Case Study for Wildfires

The proposed methodology was implemented in a case study utilizing the IEEE 30-bus test system, specifically mapped onto a region in the Yosemite-Ritter section of the Sierra Nevada in California as shown in Figure 6.5. The study area, situated between latitudes 37.6° and 38.1° and longitudes −120.7° and −120°, spans approximately 820,000 acres (331,840 hectares). This region encompasses diverse barriers and terrain types, lakes, forests, and rural areas. This selection was deliberate, as it allows for a comprehensive evaluation of the methodology's effectiveness in assessing the potential impacts of grid-induced wildfires in

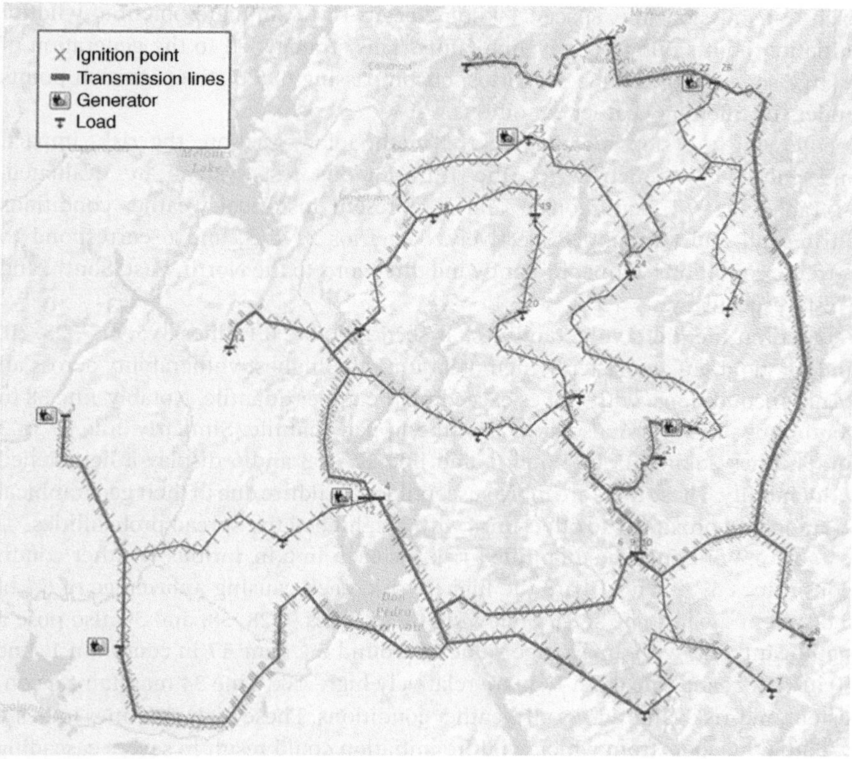

Figure 6.5 Geographic mapping of the IEEE 30-bus test system in the selected region.

regions possessing susceptible landscape characteristics. By including different barrier types and terrain features, the case study provides insights into grid resilience in the face of wildfires, considering the challenges posed by different geographical conditions.

Landscape and fuel data were extracted from LANDFIRE [37], a national initiative committed to providing comprehensive geospatial data. LANDFIRE's scope covers vegetation, wildland fuel, and fire regimes across the entirety of the United States, making it a valuable resource for this analysis. Two types of weather datasets are employed in constructing wildfire scenarios. The first set includes two historical weather datasets corresponding to distinct seasonal conditions—one during summer and another during fall. The second set comprised four synthetically generated datasets, each emulating severe extreme weather conditions but with different wind directions. This holistic approach ensures that the scenarios cover various weather conditions, from baseline weather patterns to more extreme situations. The scenarios further included strategically placed ignition points on each transmission line, spaced 1 kilometer apart, to simulate potential wildfire initiation points due to power line faults. This strategy led to the generation of a large set of 4116 unique scenarios, encompassing 686 distinct ignition points under six different weather conditions.

Following the steps detailed in the methodology section, the risk, impact, and vulnerability metrics for the IEEE 30-bus test system are evaluated. In Tables 6.1–6.4, conditions 1 and 2 represent historical weather conditions during fall and summer, respectively. Scenarios 3, 4, 5, and 6 correspond to extreme conditions with dominant wind directions to the North, East, South, and West, respectively.

Based on the line's vulnerabilities, as seen in Table 6.1, lines 3, 5, 6, 7, 28, 30, and 41 stand out, consistently demonstrating the highest vulnerability across all weather conditions, with values exceeding the upper quantile. Notably, line 38 in conditions 3, 5, and 6 also surpasses the upper quantile. Similarly, line 27 in 3 and 4, lines 26 and 25 in 4 and 6, and line 17 in 3 and 6 display a heightened vulnerability. These lines are more susceptible to wildfire due to their geographical location and proximity to other lines with heightened fire spread probabilities.

Table 6.2 presents the quantified risk for each line in various weather conditions. Lines 37, 8, and 10 are the highest risk, each causing an average of 6% of the system's load to be shed. Following these, lines 9, 28, 38, and 39 also pose a significant risk, with an average value of around 5%. Line 17 in condition 1, line 20 in 1, 4, 5, and line 40 in 1, 3, are relatively high-risk. Line 34 maintains a consistent mid-risk value across all weather conditions. These high-risk lines indicate potential hotspots from which wildfire initiation could result in severe cascading impacts on other lines and nodes due to their strategic location and surrounding available fuel conditions.

Table 6.3 provides the vulnerability metrics for the nodes under the given scenarios. Nodes 10, 13, 14, and 15 exhibit a high vulnerability of 0.48, implying they are directly impacted in nearly half of the scenarios. Nodes 2, 3, 8, and 16 also demonstrate significant vulnerability, with a value of 0.29. The high vulnerability of these nodes stems from their interconnections and dependencies on multiple lines, which increase their susceptibility to the impacts of wildfires.

The impact metric in Table 6.4 illustrates the nodes that experience the most significant outages due to wildfires. Node 30 registers the highest average impact (0.064 for all weather conditions), followed by nodes 27 and 29 (0.036 and 0.038 on average, respectively). These nodes represent the most vulnerable nodes in the system and are strong candidates for reinforcement strategies, such as integration with microgrids.

Figures 6.6 and 6.7 offer a geographical representation of the metrics' mean values across various weather conditions. By integrating the data from the tables into a visual format, the spatial representation highlights areas most susceptible to wildfires and identifies critical infrastructure that necessitates retrofitting and

Table 6.1 Assessed values for vulnerability of lines.

Line #	From node	To node	Rate (MVA)	Line's vulnerability in different weather conditions						
				1	2	3	4	5	6	Mean
1	1	2	130	0.013	0.013	0.024	0.024	0.015	0.027	0.019
2	1	3	130	0.008	0.012	0.017	0.018	0.020	0.020	0.016
3	2	4	65	**0.056**	**0.072**	**0.086**	**0.080**	**0.070**	**0.066**	**0.072**
4	3	4	130	0.016	0.018	0.013	0.015	0.009	0.009	0.013
5	2	5	130	**0.069**	**0.066**	**0.068**	**0.069**	**0.086**	**0.063**	**0.070**
6	2	6	65	**0.081**	**0.090**	**0.109**	**0.100**	**0.101**	**0.101**	**0.097**
7	4	6	90	**0.057**	**0.057**	**0.045**	**0.053**	**0.047**	0.033	**0.049**
8	5	7	70	0.018	0.018	0.012	0.012	0.012	0.015	0.014
9	6	7	130	0.022	0.024	0.028	0.034	0.021	0.027	0.026
10	6	8	32	0.021	0.015	0.017	0.012	0.018	0.009	0.015
17	12	14	32	0.027	0.030	**0.042**	0.036	0.038	**0.053**	0.038
18	12	15	32	0.034	0.036	0.042	0.024	0.036	0.042	0.036
19	12	16	32	0.039	0.029	0.041	0.032	0.038	0.042	0.037
20	14	15	16	0.024	0.013	0.019	0.015	0.013	0.021	0.018
21	16	17	16	0.015	0.028	0.025	0.018	0.018	0.016	0.020
22	15	18	16	0.022	0.019	0.016	0.030	0.019	0.009	0.019
23	18	19	16	0.001	0	0	0	0.004	0.000	0.001
24	19	20	32	0	0.003	0.003	0	0.004	0.006	0.003
25	10	20	32	0.026	0.041	0.041	**0.042**	0.029	**0.042**	0.037
26	10	17	32	0.027	0.030	0.040	**0.046**	0.040	**0.045**	0.038
27	10	21	32	0.039	0.040	**0.053**	**0.043**	0.037	0.036	0.041
28	10	22	32	**0.057**	**0.057**	**0.052**	**0.055**	0.039	**0.054**	**0.052**
29	21	22	32	0.013	0.012	0.015	0.010	0.019	0.013	0.014
30	15	23	16	0.014	0.015	0.015	0.014	0.014	0.012	0014
31	22	24	16	0.006	0.003	0.011	0.008	0.008	0.005	0.007
32	23	24	16	0.004	0.001	0.001	0.001	0.006	0.007	0.004
33	24	25	16	0	0.004	0	0	0	0.003	0.002
34	25	26	16	0.025	0.010	0.012	0.013	0.012	0.013	0.014
35	25	27	16	0.013	0.016	0.013	0.007	0.012	0.013	0.013
37	27	29	16	0.025	0.028	0.041	0.031	0.024	0.034	0.031
38	27	30	16	0.035	0.041	**0.047**	0.041	**0.044**	**0.044**	0.042
39	29	30	16	**0.045**	**0.046**	**0.049**	**0.045**	**0.045**	**0.048**	**0.046**
40	8	28	32	0.014	0.025	0.025	0.028	0.017	0.025	0.022
41	6	28	32	**0.084**	**0.083**	**0.076**	**0.081**	**0.089**	**0.083**	**0.083**

Table 6.2 Assessed values for risk of lines.

Line #	From node	To node	Rate (MVA)	Line's risk in different weather conditions 1	2	3	4	5	6	Mean
1	1	2	130	0	0	0	0	0	0	0
2	1	3	130	0.001	0.002	0	0.001	0.001	0.001	0.001
3	2	4	65	0	0.001	0	0.001	0.002	0.001	0.001
4	3	4	130	0	0	0	0	0	0.005	0.001
5	2	5	130	0	0	0.005	0	0	0	0.001
6	2	6	65	0.002	0	0.002	0	0.004	0	0.002
7	4	6	90	0	0.005	0.005	0.005	0	0	0.002
8	5	7	70	**0.057**	**0.064**	**0.057**	**0.085**	**0.049**	**0.078**	**0.065**
9	6	7	130	**0.064**	**0.056**	**0.056**	**0.040**	**0.056**	**0.080**	**0.059**
10	6	8	32	**0.038**	**0.068**	**0.060**	**0.053**	**0.068**	**0.105**	**0.065**
17	12	14	32	**0.020**	0.013	0.016	0.014	0.013	0.016	0.015
18	12	15	32	0.002	0	0.002	0	0	0	0.001
19	12	16	32	0.008	0.016	0.011	0.008	0.009	0.007	0.010
20	14	15	16	**0.020**	0.010	0.010	**0.020**	**0.023**	0.015	0.016
21	16	17	16	0	0.007	0.005	0.011	0	0	0.004
22	15	18	16	0	0	0	0	0	0	0
23	18	19	16	0.006	0	0	0.004	0	0	0.002
24	19	20	32	0	0.002	0	0	0.003	0	0.001
25	10	20	32	0	0.003	0.005	0.006	0.006	0.005	0.004
26	10	17	32	0	0	0.003	0	0	0	0.001
27	10	21	32	0	0	0.017	0.017	0.008	0.008	0.008
28	10	22	32	**0.031**	**0.031**	**0.070**	**0.046**	**0.061**	**0.070**	**0.052**
29	21	22	32	0	0	0	0	0	0	0
30	15	23	16	0.001	0.001	0	0	0.002	0	0.001
31	22	24		0	0	0	0	0	0	0
32	23	24	16	0.003	0.005	0	0	0	0	0.001
33	24	25	16	0.002	0.005	0.010	0.005	0	0.002	0.004
34	25	26	16	0.018	**0.018**	0.018	**0.018**	**0.018**	0.018	0.018
35	25	27	16	0.004	0	0	0.002	0.002	0.004	0.002
37	27	29	16	**0.067**	**0.066**	**0.069**	**0.072**	**0.072**	**0.069**	**0.069**
38	27	30	16	**0.054**	**0.057**	**0.056**	**0.054**	**0.054**	**0.059**	**0.056**
39	29	30	16	**0.047**	**0.056**	**0.056**	**0.057**	**0.057**	**0.056**	**0.055**
40	8	28	32	**0.023**	0.014	**0.024**	0.007	0.001	0.009	0.013
41	6	28	32	0.003	0	0.001	0.001	0.001	0	0.001

Table 6.3 Assessed values for vulnerability of nodes.

Node #	Demand (MW)	Node vulnerability in different weather conditions						
		1	2	3	4	5	6	Mean
1	0	0.064	0.064	0.074	0.071	0.067	0.074	0.069
2	21.7	0.281	**0.286**	**0.294**	0.284	**0.294**	**0.305**	**0.291**
3	0	**0.293**	**0.299**	**0.296**	**0.306**	**0.299**	0.284	**0.296**
4	0	0.112	0.115	0.114	0.115	0.098	0.109	0.111
5	3.2	0.067	0.067	0.069	0.067	0.071	0.071	0.069
6	0	0.238	0.259	0.262	0.255	0.246	0.245	0.251
7	3	0.069	0.071	0.076	0.077	0.071	0.069	0.072
8	7.6	**0.293**	**0.299**	**0.296**	**0.306**	**0.299**	0.284	**0.296**
9	0	0.172	0.168	0.169	0.168	0.185	0.166	0.171
10	0	**0.483**	**0.477**	**0.469**	**0.494**	**0.483**	**0.483**	**0.481**
11	22.8	0.061	0.063	0.063	0.066	0.057	0.057	0.061
12	30	0.114	0.115	0.115	0.121	0.117	0.105	0.114
13	0	**0.483**	**0.477**	**0.469**	**0.494**	**0.483**	**0.483**	**0.481**
14	5.8	**0.483**	**0.477**	**0.469**	**0.494**	**0.483**	**0.483**	**0.481**
15	0	**0.483**	**0.477**	**0.469**	**0.494**	**0.483**	**0.483**	**0.481**
16	11.2	**0.293**	**0.299**	**0.296**	**0.306**	**0.299**	0.284	**0.296**
17	6.2	0.069	0.076	0.089	0.076	0.076	0.099	0.081
18	8.2	0.134	0.131	0.136	0.128	0.130	0.134	0.132
19	3.5	0.083	0.071	0.089	0.079	0.083	0.089	0.082
20	9	0.076	0.089	0.093	0.090	0.089	0.092	0.088
21	3.2	0.051	0.048	0.047	0.055	0.048	0.038	0.048
22	9.5	0.025	0.026	0.028	0.023	0.032	0.029	0.027
23	2.2	0.070	0.082	0.085	0.085	0.069	0.087	0.079
24	17.5	0.070	0.070	0.082	0.073	0.066	0.064	0.071
25	8.7	0.071	0.070	0.076	0.071	0.076	0.074	0.073
26	0	0.077	0.069	0.069	0.064	0.069	0.070	0.069
27	3.5	0.047	0.032	0.034	0.035	0.034	0.035	0.036
28	0	0.238	0.259	0.262	0.255	0.246	0.245	0.251
29	2.4	0.069	0.071	0.086	0.070	0.064	0.077	0.073
30	10.6	0.071	0.074	0.079	0.073	0.076	0.076	0.075

Table 6.4 Assessed values for the impact of nodes.

Node #	Demand (MW)	Node's impact in different weather conditions						
		1	2	3	4	5	6	Mean
1	0	0	0	0	0	0	0	0
2	21.7	0	0	0.005	0.001	0.001	0.001	0.001
3	0	0	0	0	0	0	0	0
4	0	0	0	0	0	0	0	0
5	3.2	0.003	0.002	0.001	0.001	0	0.001	0.001
6	0	0	0	0	0	0	0	0
7	3	0.002	0.004	0.001	0.002	0.003	0.007	0.003
8	7.6	0	0	0.001	0.001	0	0.001	0
9	0	0	0	0	0	0	0	0
10	0	0	0	0	0	0	0	0
11	22.8	**0.025**	**0.025**	**0.024**	**0.027**	**0.022**	**0.031**	**0.026**
12	30	**0.016**	**0.019**	**0.021**	**0.014**	**0.013**	**0.024**	**0.018**
13	0	0	0	0	0	0	0	0
14	5.8	0	0	0.001	0.001	0	0.001	0.001
15	0	0	0	0	0	0	0	0
16	11.2	0	0	0.001	0.001	0	0.001	0
17	6.2	**0.034**	**0.019**	**0.024**	**0.027**	**0.027**	**0.026**	**0.026**
18	8.2	0	0	0.001	0.001	0	0.001	0
19	3.5	**0.013**	**0.028**	**0.020**	**0.014**	**0.015**	**0.013**	**0.017**
20	9	0	0.003	0.007	0.008	0.003	0.004	0.004
21	3.2	0.002	0	0.001	0.004	0.005	0.001	0.002
22	9.5	0.001	0	0.001	0.001	0	0.001	0.001
23	2.2	0	0.005	0.002	0.001	0.008	0.004	0.003
24	17.5	0.007	0.007	**0.012**	0.007	**0.016**	**0.011**	**0.010**
25	8.7	0	0.001	0.001	0.001	0	0.001	0.001
26	0	0	0	0	0	0	0	0
27	3.5	**0.047**	**0.032**	**0.034**	**0.036**	**0.034**	**0.036**	**0.036**
28	0	0	0	0	0	0	0	0
29	2.4	**0.035**	**0.037**	**0.039**	**0.040**	**0.038**	**0.039**	**0.038**
30	10.6	**0.058**	**0.063**	**0.068**	**0.063**	**0.063**	**0.066**	**0.064**

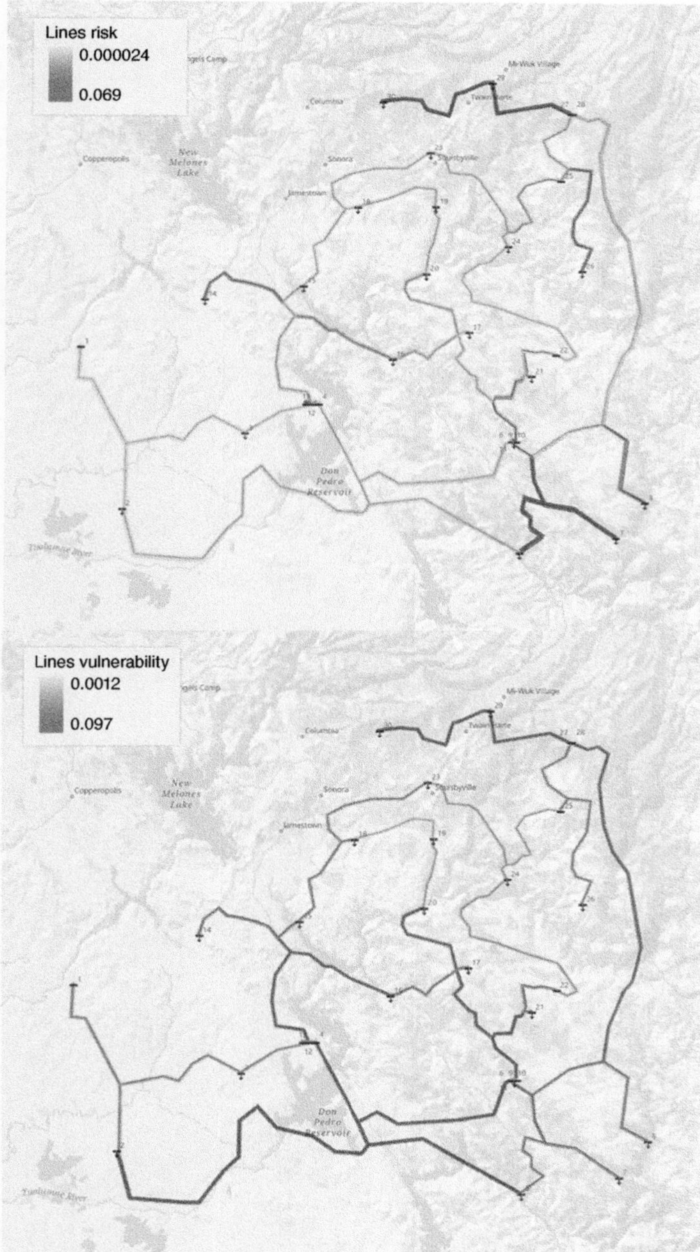

Figure 6.6 Visualization of average vulnerabilities and risk of lines.

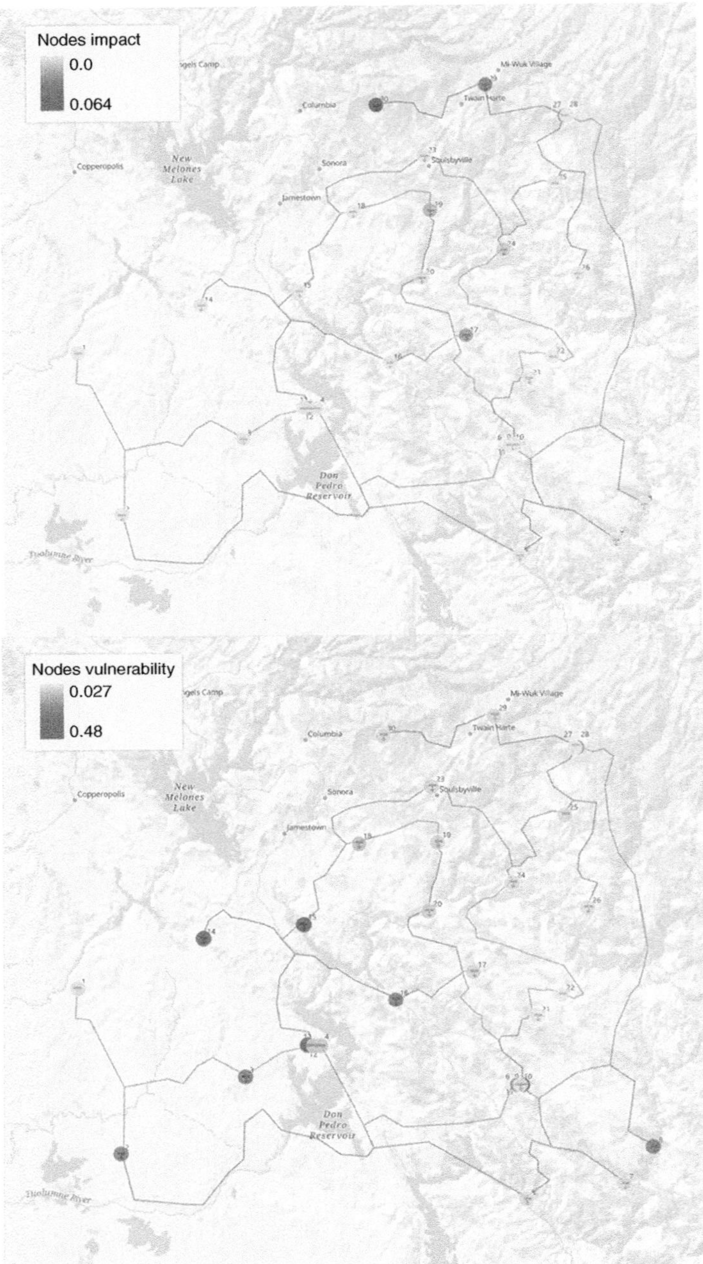

Figure 6.7 Visualization of average vulnerabilities and impact of nodes.

reinforcement. This holistic view provides stakeholders with a clear context, enabling them to grasp the broader implications. In turn, they can strategically allocate resources, implement mitigation measures, and devise preventive actions. Aligning strategies with zones of heightened vulnerability, risk, and impact reduces the potential for grid-induced wildfires and paves the way for a more resilient power infrastructure.

The case study results provide a comprehensive insight into the risk and vulnerability characteristics of the lines and nodes within the 30-bus test system. This analysis identifies components with consistently high vulnerability across different weather scenarios. Especially during the summer period, represented by condition 2, the grid is at an increased risk due to the season's pronounced wildfire potential. The combination of elevated temperatures and decreased humidity creates drier landscapes, making them more susceptible to both ignition from potential grid-related sparks and the rapid spread of any subsequent wildfires. This seasonal factor amplifies the inherent vulnerabilities of certain lines and nodes. Moreover, prevailing winds in conditions 3 and 6, directed toward the north and west, respectively, intensify the impact on specific components. These effects arise from design vulnerabilities, as well as interactions with regional topographical and vegetational factors. On the other hand, while condition 5, associated with wind directed to the south, may have a milder direct impact on most components, lines 6 and 5 display the highest vulnerability compared to other wind conditions. Thus, this condition should not be overlooked. Even minimal disruptions in such scenarios can result in cascading effects within a tightly interconnected grid. In general, lines 3, 5, 6, 7, 28, 30, 41, and nodes 10, 13, 14, and 15 emerge as the most susceptible to wildfires, attributed to their strategic positions and interconnections within the grid in all weather conditions. The failure of these components could dramatically propagate the wildfire impact across the network, leading to considerable load shedding. As a direct implication, these high-risk components should be targeted for risk mitigation strategies to reduce the overall system vulnerability. Conversely, nodes such as 30, 27, 29, 11, 17, 12, and 19, which register high outage rates during wildfires, should be prioritized for reinforcement measures. Improving the resilience of these nodes could significantly enhance the system's capacity to withstand and recover from wildfire-induced damages. This analysis offers a roadmap for power system planners and operators, anchoring the design of apt resilience and mitigation methodologies. Taking cues from the derived insights, strategies might encompass the deployment of microgrids and virtual power plants. Additionally, vegetation management, line undergrounding, and proactive asset upgrades and maintenance emerge as other viable countermeasures. The case study demonstrates the proposed methodology's effectiveness in assessing the power grid components' risk, impact, and vulnerability to wildfires. By identifying high-risk lines and nodes, the study enables informed decision-making on where to concentrate investments for resilience and risk mitigation strategies. Future research could explore more nuanced measures of risk and vulnerability and assess various mitigation and reinforcement strategies' effectiveness.

6.5 Summary

This chapter proposed a wildfire risk rating framework integrating a wildfire spread simulator with power flow analysis, landscape data, and grid topology. The proposed framework quantifies the risk emanating from power lines and the subsequent vulnerability of interconnected components due to grid-ignited wildfires. The methodology evaluates the susceptibility of the grid to wildfires, measures the expected consequences to the grid due to the failure of individual transmission lines, and assesses the average power outages at each node for a range of wildfire scenarios. The 30-bus test system was utilized for a case study to illustrate the effectiveness of the proposed model. The numerical analysis shows that the proposed framework in this study can serve as an efficient tool for power system planners and operators, providing risk-informed decisions for their capital programs and strategic investments in resilience building and risk mitigation efforts.

References

1 M. Mohler, "CAL FIRE investigators determine the cause of the campfire," California Dept. Forestry Fire Protection, Sacramento, CA, USA, Tech. Rep., 2019.

2 I. Penn, "PG&E says wildfire victims back settlement in bankruptcy," *The New York Times*, May 2020.

3 "CAL FIRE investigators determine cause of the Dixie fire." California Department of Forestry and Fire Protection, 2021. Accessed: May 2023. [Online]. Available: https://www.fire.ca.gov/media/edwez51p/dixie_fire_release.pdf

4 "PG&E responds to CAL FIRE's report on 2021 Dixie fire in Butte, Lassen, Plumas, Shasta, and Tehama Counties." July 2021. Accessed: May 2023. [Online]. Available: https://s1.q4cdn.com/880135780/files/doc_downloads/2022/06/DIXIE_FINAL_Redacted_PGE-Proposed-Redactions-Statement-49411384.pdf

5 K. C. Short, "Spatial wildfire occurrence data for the United States, 1992–2020 [FPA_FOD_20221014]." 6th ed. Fort Collins, CO, USA: Forest Service Research Data Archive, 2022. Accessed: May 2023. [Online]. Available: https://doi.org/10.2737/RDS-2013-0009.6.

6 W. Jahn, J. L. Urban, and G. Rein, "Powerlines and wildfires: overview, perspectives, and climate change: could there be more electricity blackouts in the future?," *IEEE Power Energy Mag.*, vol. 20, no. 1, pp. 16–27, Jan. 2022.

7 B. Chiu, R. Roy, and T. Tran, "Wildfire resiliency: California case for change," *IEEE Power Energy Mag.*, vol. 20, no. 1, pp. 28–37, Jan. 2022.

8 E. A. Udren, C. Bolton, D. Dietmeyer, T. Rahman, and S. Flores-Castro, "Managing wildfire risks: Protection system technical developments combined with operational advances to improve public safety," *IEEE Power Energy Mag.*, vol. 20, no. 1, pp. 64–77, Jan. 2022.

9 D. Sharafi *et al.*, "Wildfires down under: Impacts and mitigation strategies for Australian electricity grids," *IEEE Power Energy Mag.*, vol. 20, no. 1, pp. 52–63, Jan. 2022.

10 M. Davoudi, B. Efaw, M. Avendaño-Mora, J. L. Lauletta, and G. B. Huffman, "Reclosing of distribution systems for wildfire prevention," *IEEE Trans. Power Deliv.*, vol. 36, no. 4, pp. 2298–2307, Jan. 2021.

11 S. Taylor and L. A. Roald, "A framework for risk assessment and optimal line upgrade selection to mitigate wildfire risk," *Electr. Power Syst. Res.*, vol. 213, pp. 108592, Dec. 2022.

12 S. Ghosh and S. Dutta, "A comprehensive forecasting, risk modelling and optimization framework for electric grid hardening and wildfire prevention in the US," *Int. J. Energy Eng.*, vol. 10, no. 3, pp. 80–89, 2020.

13 R. Bayani, M. Waseem, S. D. Manshadi, and H. Davani, "Quantifying the risk of wildfire ignition by power lines under extreme weather conditions," *IEEE Syst. J.*, vol. 17, no. 1, pp. 1024–1034, Jul. 2022.

14 N. Rhodes and L. A. Roald, "Co-optimization of power line shutoff and restoration under high wildfire ignition risk," in *2023 IEEE Belgrade PowerTech*, Jun. 2023, pp. 1–7.

15 A. Kody, R. Piansky, and D. K. Molzahn, "Optimizing transmission infrastructure investments to support line de-energization for mitigating wildfire ignition risk," arXiv preprint, arXiv:2203.10176, Mar. 2022.

16 J. M. Johnson, "Quantifying the economic risk of wildfires and power lines in San Diego County," Duke Univ., Durham, NC, USA, May 2014.

17 B. D. Russell, J. A. Wischkaemper, C. L. Benner, and K. Manivannan, "Preventing certain powerline caused wildfires by early detection and repair of failing devices," in *2022 IEEE/PES Transmission Distrib. Conf. Exp. (T&D)*, Apr. 25, 2022, pp. 1–5.

18 J. Ma, J. C. Cheng, F. Jiang, V. J. Gan, M. Wang, and C. Zhai, "Real-time detection of wildfire risk caused by powerline vegetation faults using advanced machine learning techniques," *Adv. Eng. Inform.*, vol. 44, p. 101070, Apr. 2020.

19 M. Zhao and M. Barati, "A real-time fault localization in power distribution grid for wildfire detection through deep convolutional neural networks," *IEEE Trans. Ind. Appl.*, vol. 57, no. 4, pp. 4316–4326, May 2021.

20 S. Kandanaarachchi, N. Anantharama, and M. A. Munoz, "Early detection of vegetation ignition due to powerline faults," *IEEE Trans. Power Deliv.*, vol. 36, no. 3, pp. 1324–1334, Jul. 2020.

21 H. Nazaripouya, "Power grid resilience under wildfire: A review on challenges and solutions," in *2020 IEEE Power Energy Soc. Gen. Meeting (PESGM)*, Aug. 2020, pp. 1–5.

22 M. Abdelmalak and M. Benidris, "Enhancing power system operational resilience against wildfires," *IEEE Trans. Ind. Appl.*, vol. 58, no. 2, pp. 1611–1621, Jan. 2022.

23 D. L. Donaldson, M. S. Alvarez-Alvarado, and D. Jayaweera, "Power system resiliency during wildfires under increasing penetration of electric vehicles," in *2020 Int. Conf. Prob. Met. Appl. to Power Syst. (PMAPS)*, Aug. 2020, pp. 1–6.

24 R. Bayani and S. D. Manshadi, "Resilient expansion planning of electricity grid under prolonged wildfire risk," *IEEE Trans. Smart Grid*, vol. 13, no. 1, pp. 1–10, Jan. 2023.

25 W. Yang, S. N. Sparrow, M. Ashtine, D. C. Wallom, and T. Morstyn, "Resilient by design: Preventing wildfires and blackouts with microgrids," *Appl. Energy*, vol. 313, pp. 118793, May 2022.

26 R. Moreno *et al.*, "Microgrids against wildfires: Distributed energy resources enhance system resilience," *IEEE Power Energy Mag.*, vol. 20, no. 1, pp. 78–89, Jan. 2022.

27 Q. Al Saeed and H. Nazaripouya, "Impact of wildfires on power systems," in *2022 IEEE Int. Conf. Environ. Electr. Eng. 2022 IEEE Ind. Commercial Power Syst. Europe (EEEIC/I&CPS Europe)*, Jun. 2022, pp. 1–5.

28 M. Choobineh, B. Ansari, and S. Mohagheghi, "Vulnerability assessment of the power grid against progressing wildfires," *Fire Saf. J.*, vol. 73, pp. 20–28, Apr. 2015.

29 A. Malik *et al.*, "Data-driven wildfire risk prediction in northern California," *Atmos.*, vol. 12, no. 1, p. 109, Jan. 2021.

30 J. W. Mitchell, "Power line failures and catastrophic wildfires under extreme weather conditions," *Eng. Fail. Anal.*, vol. 35, pp. 726–735, Dec. 2013.

31 D. A. Dawe *et al.*, "Short-and long-term wildfire threat when adapting infrastructure for wildlife conservation in the boreal forest," *Ecol. Appl.*, vol. 32, no. 6, pp. e2606, Sep. 2022.

32 R. Zhang et al., "Risk assessment of high voltage power lines crossing forest areas–a case study of wildfires," *ISPRS Ann. Photogramm. Remote Sens. Spat. Inf. Sci.*, vol. 10, pp. 199–206, Oct. 2022.

33 G. D. Papadopoulos and F. N. Pavlidou, "A comparative review on wildfire simulators," *IEEE Syst. J.*, vol. 5, no. 2, pp. 233–243, Jun. 2011.

34 M. A. Finney, "FARSITE: Fire area simulator-model development and evaluation," U.S. Department of Agriculture, Forest Service, Rocky Mountain Research Station, Ogden, UT, USA, 1988.

35 M. P. Nicolau et al., "Large forest fire risk assessment and fuel management: Operational tools and integrated approach," in *Fire Efficient, Operational Tools for Improving Efficiency in Wildfire Risk Reduction in EU Landscapes*. European Union, Sep. 2014.

36 P. N. Yasasvi, S. Chandra, A. Mohapatra, and S. C. Srivastava, "An exact SOCP formulation for AC optimal power flow," in *2018 20th Nat. Power Syst. Conf. (NPSC)*, IEEE, Dec. 2018, pp. 1–6.

37 M. G. Rollins, "LANDFIRE: A nationally consistent vegetation, wildland fire, and fuel assessment," *Int. J. Wildland Fire*, vol. 18, no. 3, pp. 235–249, 2009.

7

The Economics of Pre-Disaster Resource Mobilization

7.1 Principles of Resource Mobilization

Climate shocks and natural disasters, in particular hurricanes, can result in significant economic, social, and physical disruptions and cause considerable inconvenience for residents living in disaster areas due to loss of electricity, water, and communication [1]. Therefore, a comprehensive study of this issue from different perspectives is needed to find efficient ways of improving the resilience of these critical lifeline systems.

Various studies have been proposed in the literature in the context of emergency planning for power systems. In [2], the research problems and models for substations and/or distribution feeders planning under normal and emergency conditions were reviewed and discussed. A case study on hurricane planning and rebuilding the electrical infrastructure along the Gulf Coast region of the United States for Hurricane Katrina was presented in [3]. A risk assessment method for infrastructure technology planning to improve the power supply resiliency to natural disasters was proposed in [4]. Reduced cost, as well as power supply availability, were considered two fundamental decision factors in their hurricane planning approach. In [5], a stochastic integer program was proposed to find the optimal schedule for inspection, damage evaluation, and repair of post-earthquake damaged electric power systems. The aim was to minimize the average time that each customer was without power. A comprehensive survey of models and algorithms for emergency response logistics in electric distribution systems, including reliability planning with fault considerations and contingency planning models, was presented in [6] and [7].

In the context of physical behavior analysis of power system infrastructure in hurricane disasters, [8] analyzed the resilience of power distribution systems based on the power distribution infrastructure and its interaction with the biophysical environment, as well as how restoration processes are prioritized. It

The Economics of Climate Resilience in Power Infrastructure, First Edition. Ali Arabnya and Amin Khodaei.

was concluded that even though the infrastructure does not significantly affect outage duration, the interaction between infrastructure and the biophysical environment significantly affects the outage duration. In another study, a data mining approach was proposed to evaluate the impact of soil and topographic variables on the accuracy of the power outage prediction models in hurricane events [9]. The results indicate that certain land cover variables can be approximated for the power system and incorporated into the model when detailed information about the power system is unavailable. In [10], a method for the characterization of the behavior of networked infrastructure, including power delivery systems in natural hazard events such as hurricanes, was presented. The model also included resilience and interdependency measures. The model can be utilized to develop design strategies for improved power infrastructure resiliency in natural disasters. Reference [11] proposed a probabilistic framework for vulnerability analysis of distribution poles subject to hurricane hazards considering the impact of changing climate. The results indicate that changing climate and the age of the poles significantly increase the failure rate of distribution poles.

Outage prediction is an important means of obtaining an efficient response to a hurricane. In this context, [12] introduced a method for estimating the restoration time of electric power systems after hurricanes and ice storms. Using a large dataset of six hurricanes and eight ice storms, the accelerated failure time models were developed to forecast the duration of each probable outage. In [13], the negative binomial regression models for predicting outages due to hurricanes were developed. The number of transformers in the area, maximum wind gust speed, the power company affected, and the hurricane effect turned out to be the most explanatory variables. Diagnostic statistics such as pseudo-R-squared values were used for model selection purposes. Their adopted zip code-based model can be used to predict the likely outage rates prior to hurricane events. In another work, [14] used regression analysis and data mining to develop models to estimate the number of utility poles that will be damaged based on damage data from past storms. Results indicate that hurricane-related damages to the poles can be predicted accurately, given that past damage data are available and adequate. However, the availability of past data can be challenging, limiting the model's efficiency in practice.

In the context of resource allocation for the restoration of power systems, [15] presented three mathematical goal programming models to locate the repair units and restore the transmission and distribution lines efficiently. The first model finds the optimal repair-unit dispatch tactical plan with a forecast of adverse weather conditions. The second model derives the optimal repair-unit location for a short-term strategic plan under normal weather conditions. The third model finds the optimal number of repair units for a long-term strategic plan. In another work, a mixed-integer programming model and a general column-generation approach were proposed for inventory decision-making of power system components throughout a populated area to maximize the amount of power served after disaster restoration [16]. In [17], service restoration, considering the restrictions on emergency-response logistics, was studied to minimize customer interruption costs. The reconfiguration and the resource dispatching issues were considered systematically to derive the optimal time sequence for every step of the restoration plan. In [18], a decision-making model was proposed to manage the required resources for economic power restoration operations. The optimal number of depots, the optimal location of depots, and the optimal number of repair crews were determined by their model to minimize the transportation cost associated with restoration operations. In [19], a decision support tool for improving information used by electric utilities to manage the restoration of power distribution components damaged due to large-scale storms was described. The circuit layout, the placement of protective and switching devices, and the location of customers were taken into account to allocate the crew resources to manage the storm outage in a cost-effective manner.

Ref. [20] introduced a two-stage optimization model to enhance the resilience of power systems against extreme cold events. The first stage involves optimizing the output of generation units to improve system stability and load margin under varying fault scenarios, aiming to mitigate the rapid increase in load demand during extreme cold events. The second stage includes an optimal allocation of emergency resources, mobile energy storage, and critical line inspections, to minimize economic losses due to equipment failures and load curtailments. Authors in [21] investigated the use of mobile wind turbines to improve the resilience of microgrids during extreme events. Using a scenario-based stochastic mixed-integer linear program, an optimization model for the pre-positioning of mobile wind turbines was presented. The proposed model accounts for uncertainties in road conditions, power line faults, and wind energy generation forecasts. A case study involving a central Alabama interstate transportation network and the IEEE 33-node test power systems was conducted to demonstrate the effectiveness of the proposed model. A two-stage stochastic programming model, designed to improve the resilience of power distribution systems in the face of natural disasters, was presented in [22]. The first stage involves creating a wind field model to predict the wind speed and a stress-strength model to evaluate the likelihood of distribution line failure to strategically position repair crew and mobile emergency generators. The second-stage coordinates network reconfiguration, dispatches repair crews, and mobilizes mobile emergency generators to minimize load shedding and expedite repairs following a disaster.

Although a variety of problems for power system planning in hurricane-prone areas have been addressed in the literature, to the best of our knowledge, a few provide a comprehensive and generic approach to resource allocation. This chapter develops an efficient decision-making tool for proactive restoration planning of power systems to minimize the expected customer load interruption, restoration operation, and electricity generation costs. The physics and the economy of the power system, specifically the unit commitment problem, are incorporated into our model—which results in higher practicality of the proposed model for utility companies.

The rest of this chapter is organized as follows: Section 7.2 describes the model, and Section 7.3 presents the problem formulation and methodology. Section 7.4 illustrates the numerical results on the IEEE 118-bus test system. Finally, Section 7.5 provides the concluding remarks and a summary of this chapter.

7.2 Economic Variables in Resource Mobilization

Consider a power system with some of its components, including generation units, transmission lines, and substations along with downstream distribution lines, located on the path of an upcoming hurricane. The objective is to proactively allocate and mobilize available resources to enable the quick response capability of system operators to repair and restore potential damages in a way that minimizes the expected incurred costs. Furthermore, estimating additional resources that need to be outsourced to cope with the aftermath of the hurricane is another critical issue for system operators. The expected incurred cost comprises the customers' load interruption, electricity generation, and system repair and restoration costs.

The damage state of components is represented by Bernoulli random variables and is considered to have two states: *damaged* and *functional*. If, after the upcoming hurricane, the component is still *functional*, the value of 1 will be assigned, and if the component is *damaged*, this value will be 0. Weather-related failure rate and probability of component damages follow the corresponding models in [1].

The time to repair each potentially damaged component is considered stochastic and is modeled by a random variable that may take various probability distributions. Without loss of generality, it is assumed that the time to repair random variables to be defined by the Weibull density function as follows:

$$F_{T_k}(t) = \begin{cases} \dfrac{\rho}{\lambda_k}\left(\dfrac{t}{\lambda_k}\right)^{\rho-1} e^{-(t/\lambda_k)^{\rho}} & if\, t \geq 0, \\ 0 & otherwise. \end{cases} \tag{7.1}$$

where

ρ is the shape parameter

λ_k is the scale parameter
$k \in \{b, i, l\}$ in which b, i, l are indices for buses, generation units, and transmission lines, respectively

Any other appropriate probability distribution can be used in our model without loss of generality. Considering the stochastic nature of the pre-hurricane problem, resources need to be allocated to minimize the expected cost of system restoration. However, at the time of resource allocation, the damage state and the time to repair components are not yet realized. Moreover, a set of decision variables needs to be determined in the second stage once the outcomes of the hurricane are known. Therefore, the problem structure would be a two-stage stochastic problem with *recourse*. The objective is to allocate resources to minimize the first-stage resource allocation costs and the second-stage expected *recourse* costs. In this class of problems, a set of *recourse* activities can be performed for each possible stochastic scenario to compensate for any deficiency in the first-stage decisions to avoid violation of the prevailing constraints [23].

7.3 An Economic Model for Pre-Event Resource Mobilization

The problem is formulated as a two-stage stochastic linear program with recourse. The general formulation of a two-stage stochastic linear program with recourse is as follows:

$$\begin{aligned} &\min_{x} \ z = cx + Q(x) \\ &s.t. \ \ Ax = b, \quad x \in X \end{aligned} \tag{7.2}$$

where

c is the cost vector
b is the right-hand side vector
$x = [u_{bt}{}^T, v_{lt}{}^T]^T$ is the first-stage decision variable
u_{bt} is the repair state variable of line l at time t
v_{lt} is the repair state variable of bus b at time t
A is the coefficient matrix of the first-stage variable (which in our model is the cost coefficient of the resource constraint)
Function $Q(x)$ is the second-stage value function (expected recourse cost function) defined as:

$$Q(x) = \mathbb{E}_{\xi}\left[Q(x, \xi(\omega))\right] \tag{7.3}$$

with

$$Q(x, \xi(\omega)) = \min_{x} \{q(\omega)y \mid W_y = h(\omega) - T(\omega) \quad x, y \in Y\} \tag{7.4}$$

where

$y = \left[LI_{bts}^{T}, I_{its}^{T}, P_{its}^{T}, SU_{its}^{T}, SD_{its}^{T}, X_{bts}^{+T}, X_{lts}^{+T}\right]^{T}$ is the second-stage variable of the proposed model

$q(w)$ is the recourse penalty coefficient

W is the recourse matrix (which is the coefficient matrix of the second-stage variables in the constraints)

T is the technology matrix (which is the coefficient matrix of the first-stage variables in the second-stage problem's constraints)

ξ is a random N-vector in (Ω, A, S) probability space (which in our model is the multivariate random variable of damage state and time to repair for all components under hurricane damage risk) [24].

In Eq. (7.4), LI_{bt} is the load interruption variable at bus b at time t, I_{it} is the Commitment state variable of generating unit i at time t, P_{it} is the real power generation of unit i at time t, SU_{it} is the startup cost variable of unit i at time t, and SD_{it} is the shutdown cost variable of unit i at time t.

7.3.1 Objective Function

The objective of the pre-hurricane model is to minimize the cost of the first-stage resource allocation decisions, and the expected cost of second-stage system configuration as follows:

$$\min_{u,v} \sum_{t}\sum_{b} C_{bt}R_{bt}u_{bt} + \sum_{t}\sum_{l} C_{lt}R_{lt}v_{lt} + \cdots$$

$$\mathbb{E}_S\left[\min_{LI,I,P,SU,SD,X_l^+,X_b^+} \sum_{t}\sum_{b} VOLL_{bt}LI_{bts} + \sum_{t}\sum_{i}\left(C_{it}^{g}I_{its}P_{its} + SU_{its} + SD_{its}\right)\right.$$

$$\left. + \sum_{l}\sum_{t} q_{lt}^{+}X_{lts}^{+} + \sum_{b}\sum_{t} q_{bt}^{+}X_{bts}^{+}\right] \tag{7.5}$$

The first term represents the cost of resources primarily allocated to substations (and their downstream distribution lines), where C_b is the hourly crew cost per person for bus b repair, and the second term is the cost of resources primarily allocated to transmission lines where C_l is the hourly crew cost per person for line l repair. The expected second-stage (recourse) function includes the load interruption cost over the restoration planning horizon and the total generation cost, including fuel costs, startup costs, and shutdown costs in scenario s where C_{it}^{g} is Generation cost of unit i at time t, C_{it}^{sd} is the shutdown cost of unit i at time

t, and C_{it}^{su} is the startup cost of unit i at time t. It further includes the cost of secondary resources allocated to transmission lines and substations under scenario s. These secondary resources fill the shortage of actual restoration resources that have not been allocated in the first-stage decisions but are required to accomplish the restoration operations under scenario s. Mathematically, these secondary resources eliminate the infeasibility of the second-stage problem under any decision made in the first stage.

7.3.2 Constraints

7.3.2.1 Resource Constraint

The first-stage problem is constrained by a restriction on primary resources (Eq. 7.6), representing the maximum amount of resources that can be allocated to the entire system in each time period. In Eq. (7.6), R_t is the size of the repair crew allocated to a component at time t and R_t^{max} is the number of available repair crew at time t.

$$\sum_{l} R_{lt} v_{lt} + \sum_{b} R_{bt} u_{bt} \leq R_t^{max}, \quad \forall t \tag{7.6}$$

7.3.2.2 Damage State of Generation Units

The initial damage state of each generating is represented by the outcome of random variable $\vartheta_i \sim$ Bernoulli (γ_i), where ϑ_i is the random initial state of unit i after the hurricane strikes. The damage state of generation units over the restoration horizon is modeled as follows:

$$t - My_{its} \leq \left(1 - \vartheta_i^s\right) T_i^s, \quad \forall i, \forall t, \forall s \tag{7.7}$$

$$y_{its} \leq \vartheta_i^s, \quad \forall i, \forall t = 0, \ldots, T_i, \forall s \tag{7.8}$$

where

T_i^s is the outcome of random variable $T_i \sim$ Weibull (λ_i) in scenario s
M is a large positive number
y_{iy} is the outage state variable of unit I at time t

If at the beginning of the restoration horizon of scenario s, the generation unit i is in a damaged state, then the outcome of random variable ϑ_i denoted by ϑ_i^s is 0; hence it needs to be restored. The state of a damaged generating unit does not change unless the required restoration operation is performed. On the other hand, if the generation unit is functional, then the random variable ϑ_i is 1, which indicates that it can be immediately committed for generation. It is assumed that if a generation unit is in the functional initial state, its state will remain the same up to the end of the restoration horizon for that particular scenario.

7.3.2.3 Damage State of Substations

The initial damage state of each substation b under scenario s, z_{b0s} is represented with the outcome of random variable $\phi_b \sim Bernoulli\ (\gamma_b)$. The damage state of each substation over the restoration horizon under scenario s is modeled as follows:

$$z_{b0s} = \phi_b^s, \quad \forall b, \forall s \tag{7.9}$$

$$0 \leq z_{b(t+1)s} - \left(\sum_{k=1}^{t} \left(u_{bk} + X_{bks}^+\right) - T_b^s + 0.5 \right) / M \leq 1, \quad \forall b, \forall t, \forall s \tag{7.10}$$

$$\sum_{k=t}^{t+T_b^s-1} \left(u_{bk} + X_{bks}^+\right) \geq T_b^s \left(u_{bt} + X_{bts}^+ - u_{b(t-1)} - X_{b(t-1)s}^+\right), \quad \forall b, \forall t, \forall s \tag{7.11}$$

where

T_b^s is the outcome of random variable $T_b \sim Weibull\ (\lambda_b)$

u_{bt} is the first-stage decision variable on primary resource allocation to bus b at time t; and

$$u_{bt} + X_{bts}^+ \leq 1, \quad \forall b, \forall t, \forall s, \tag{7.12}$$

Constraints (7.9) and (7.10) model the damage state of a substation during the restoration horizon, while (7.11) guarantees that enough resources are allocated to the component for repair purposes. It is assumed that if a substation under a scenario s is initially in a *functional* state ($z_{b0s} = 1$), it will remain in the same state up to the end of the restoration horizon.

7.3.2.4 Damage State of Transmission Lines

In the same way as generation units and substations, the initial damage state of transmission lines is represented by the outcome of a random variable $\psi_l \sim Bernoulli\ (\gamma_l)$. The damage state and repair duration of transmission lines in each scenario are modeled as follows:

$$\omega_{l0s} = \psi_{ls}, \quad \forall l, \forall s \tag{7.13}$$

$$0 \leq \omega_{l(t+1)s} - \left(\sum_{k=1}^{t} \left(v_{lk} + X_{lks}^+\right) - T_{ls} + 0.5 \right) \Big/ M \leq 1, \quad \forall l, \forall t, \forall s \tag{7.14}$$

$$\sum_{k=t}^{t+T_{ls}-1} \left(v_{lk} + X_{lks}^+\right) \geq T_{ls} \left(v_{lt} + X_{lt}^+ - v_{l(t-1)} - X_{l(t-1)s}^+\right), \quad \forall l, \forall t, \forall s \tag{7.15}$$

where

T_{ls} is the outcome of random variable $T_l \sim Weibull\ (\lambda_l)$

$$v_{lt} + X_{lts}^+ \leq 1, \quad \forall l, \forall t, \forall s \tag{7.16}$$

If a transmission line is damaged, its initial state variable (w_{l0s}) becomes 0 and remains unchanged until required resources are allocated and the damaged component is repaired. On the other hand, if it does not undergo any damage, w_{l0s} takes the value of 1. It is assumed that the *functional* state of a component remains unchanged during the restoration horizon.

7.3.2.5 Load Balance Constraint

The bus load balance constraint for each scenario is represented as follows:

$$\sum_{i \in N_b} P_{its} + \sum_{l \in N_b} PL_{lts} + LI_{bts} = D_{bt}, \quad \forall b, \forall t, \forall s \tag{7.17}$$

This constraint ensures that the injected power to a bus from connected transmission lines and generation units must supply the bus load in each scenario where D_{bt} is the load demand at bus b at time t, and N_b is the set of components connected to bus b; however, if the injected power is not sufficient, the load supply will be interrupted equal to the load interruption variable (LI_{bts}).

7.3.2.6 Power Generation Constraints

The real power generation of unit i is constrained by its damage state, the commitment state, and its minimum and maximum generation capacity as follows:

$$P_i^{min} y_{its} I_{its} \le P_{its} \le P_i^{max} y_{its} I_{its}, \quad \forall i, \forall t, \forall s \tag{7.18}$$

The coupling constraint of unit commitment and damage state holds for each scenario as

$$I_{its} \le y_{its}, \quad \forall i, \forall t, \forall s \tag{7.19}$$

The damage state of substations connected to each generating unit also constrains the real power generation as Eq. (7.20), where α_{ib} is the element of unit i and bus b in the generation-bus incidence matrix and z_{bt} is the outage state variable of bus b at time t.

$$-M \sum_b \alpha_{ib} z_{bts} \le P_{its} \le M \sum_b \alpha_{ib} z_{bts}, \quad \forall i, \forall t, \forall s \tag{7.20}$$

Thus, if the substation connected to a generation unit is damaged, the generating unit becomes offline.

7.3.2.7 Power Flow Constraints

The transmission network power flow is modeled as follows:

$$-PL_l^{max} w_{lts} \le PL_{lts} \le PL_l^{max} w_{lts}, \quad \forall l, \forall t, \forall s \tag{7.21}$$

$$-M \sum_b \beta_{lb}^{from} z_{bts} \le PL_{lts} \le M \sum_b \beta_{lb}^{from} z_{bts}, \quad \forall l, \forall t, \forall s \tag{7.22}$$

$$-M\sum_{b}\left|\beta_{lb}^{to}\right|z_{bts} \le PL_{lts} \le M\sum_{b}\beta_{lb}^{to}z_{bts}, \quad \forall l, \forall t, \forall s \tag{7.23}$$

$$\begin{aligned}&- M(1 - w_{lts}) - M\left(1 - \sum_{b}|\beta_{lb}|z_{bts}\right)\\ &\le PL_{lts} - \frac{\sum_{b}\beta_{lb}\delta_{bts}}{x_l}\\ &\le M(1 - w_{lts}) - M\left(1 - \sum_{b}|\beta_{lb}|z_{bts}\right), \quad \forall l, \forall b, \forall t, \forall s\end{aligned} \tag{7.24}$$

If line l at time t is in the functional state, the power can flow in either of the two directions, but not more than the maximum power flow capacity of the line (Eq. 7.21), where PL_{lt} is the power flow of line l at time t, and w_{lt} is the outage state variable of line l at time t. However, if the line is damaged, the power flow in the line will be equal to 0. In addition, as long as any of the substations connected to each particular transmission line is in the damaged state, the associated power flow is set to 0 using Eqs. (7.22)–(7.23), where β_{lb} is the element of line l and bus b in line-bus incidence matrix and δ_{bts} is the bus voltage angle. The power flow in the line is related to bus voltage angles, as in Eq. (7.24).

7.3.2.8 Unit Commitment Constraints

The unit commitment constraints for thermal generating units are an important feature of the hurricane restoration planning model. The related MIP constraints, i.e., startup and shutdown costs, ramp-up and ramp-down, and minimum uptime and downtime constraints, are imposed on the model to incorporate the impact of optimal unit commitment configuration of the system in restoration decision-making [25].

7.3.2.9 Nonanticipativity

An important issue that needs to be considered in solving stochastic programs is that the decisions should not depend on the outcome of stochastic parameters, denoted as the *nonanticipativity* concept [23]. Birge's method is one way to enforce the nonanticipativity requirement [26]. For instance, for LI_{bts} the nonanticipativity is modeled as follows:

$$\left(\sum_{s' \in S_s^t} p(s')\right) LI_{bts} = \sum_{s' \in S_s^t} p(s)\, LI_{bts'}, \quad \forall b, \forall t, \forall s \tag{7.25}$$

where

S_s^t is the set of scenarios that are identical to scenario s at time t
$p(s)$ is the probability of scenario s

A similar formulation structure for the rest of the second-stage variables is considered in the model.

7.3.3 Proposed Solution Scheme

7.3.3.1 Scenario Construction and Reduction

Due to the presence of continuous random variables, i.e., the Weibull distribution for time to repair each damaged component, the stochastic data process of the proposed models, ζ has infinite support. To make the problem tractable, the stochastic data process ζ needs to be redistributed to provide finite support with the reduced (optimal) number of scenarios. We use the Latin hypercube sampling [27] to replace ζ by a scenario tree approximation ξ_{tr}, which has a finite but large number of scenarios. The Latin hypercube sampling guarantees that the whole range of values for a random variable is sampled. For a sample size of N, the Latin hypercube sampling technique selects N different values from each random variable by dividing the range of each random variable into N non-overlapping intervals. Then, by shuffling and pairing these values, N scenarios are constructed, each with a probability of $1/N$.

The next step is to reduce the number of scenarios to a computationally tractable size. Various reduction techniques are available for different applications. For the constructed probability measure of $\xi_{tr} = \sum_{k=1}^{N}(1/N)s_k$ it is required to determine an index set $K'_* \subset \{1, \dots, N\}$ of given cardinality $\#K'_* = N - N'$ and a probability measure $\widetilde{\xi}_* = \sum_{k'=1,k'\notin K'_*}^{N} p^*_{k'}\ s_{k'}$ such that

$$\widehat{\mu}_c(\xi_{tr}, \widetilde{\zeta}_*) = inf \left\{ \widehat{\mu}_c \left(\xi_{tr}, \sum_{k=1,k\notin K'_*}^{N} p_{k'}\ s_{k'} \right) \right.$$

$$\left. : K'_* \subset \{1, \dots, N\}, \#K'_* = N - N', \sum_{k'\notin K'_*} p_{k'} = 1, \quad p_{k'} \geq 0, \ k' \notin K' \right\} \quad (7.26)$$

where *Kantorovich functional* $\widehat{\mu}_c(\xi_{tr}, \widetilde{\zeta}_*)$ is an estimation of the probability distance $\zeta_c(\xi_{tr}, \widetilde{\zeta}_*)$. Problem (7.26) can be solved through various techniques, but due to the accuracy of the backward reduction algorithm, we solve it through this method. Readers are referred to [28] for a detailed explanation of the backward scenario reduction algorithm.

7.3.3.2 Decomposition Strategy

The proposed two-stage stochastic program with recourse is converted to its deterministic mixed-integer program equivalence by constructing the scenario tree and reducing it to a tractable number of scenarios. Benders decomposition for mixed-integer programming is an efficient strategy when the original problem is large-scale and difficult to solve, while the Benders subproblem and the relaxed master problem are much more tractable to solve. To employ the decomposition strategy for the proposed problem, we consider the continuous variable vector as $X = \left[LI_{bts}^T, P_{its}^T, PL_{lts}^T, SU_{its}^T, SD_{its}^T\right]^T$, the binary variable vector as $Y = \left[u_{bt}^T, v_{lt}^T, X_{bts}^T, X_{lts}^T, y_{its}^T, z_{bts}^T, w_{lts}^T, I_{its}^T\right]^T$, and the cost coefficient matrix of the integer variables in the objective function as C^T, the cost coefficient matrix of the continuous variables in the objective function as D^T. We also assume that A and B represent the coefficient matrix of X and Y in the constraints, respectively. Finally, H is assumed to represent the right-hand side matrix of the constraints. The problem is decomposed into a master problem and a subproblem. The master problem is a pure integer program, while the subproblem is a dual linear program (without any integer variable). Considering U the dual variable vector for the subproblem, the Benders decomposition algorithm for the proposed model is shown in Algorithm 7.1. [29].

Algorithm 7.1 Benders decomposition for equivalent MIP Problem

---

{initialization}

$Y :=$ initial feasible integer solution

Lower bound $(LB) := -\infty$, Upper bound $(UB) := +\infty$

while $UB - LB > \varepsilon$ **do**

{solve dual LP subproblem}

$$max_U \left\{\sum_t \sum_b C_{bt} R_b^{min} \bar{u}_{bt} + \sum_t \sum_l C_{lt} R_l^{min} \bar{v}_{lt} + \sum_s \sum_l \sum_t p(s) q_{lt}^{+} \bar{X}_{lts}^{+} + \sum_s \sum_b \sum_t p(s) q_{bt}^{+} \bar{X}_{bts}^{+} + (H - B\bar{Y})^T U \mid A^T U \leq C, U \geq 0\right\}$$

if unbounded **then**

Get unbounded ray $\bar{U}$

Add cut $(H - BY)^T \bar{U} \leq 0$ to pure IP master problem

else

Get extreme point $\bar{U}$

Add cut $\theta \geq \sum_t \sum_b C_{bt} R_b^{min} u_{bt} + \sum_t \sum_l C_{lt} R_l^{min} v_{lt} + \sum_s \sum_l \sum_t p(s) q_{lt}^{+} \bar{X}_{lts}^{+} + \sum_s \sum_b \sum_t p(s) q_{bt}^{+} \bar{X}_{bts}^{+} + (H - BY)^T \bar{U}$ to master problem

$$UB := min \left\{UB, \sum_t \sum_b C_{bt} R_b^{min} \bar{u}_{bt} + \sum_s \sum_l \sum_t p(s) q_{lt}^{+} \bar{X}_{lts}^{+} + \sum_s \sum_b \sum_t p(s) q_{bt}^{+} \bar{X}_{bts}^{+} + \sum_t \sum_l C_{lt} R_l^{min} \bar{v}_{lt} + (H - B\bar{Y})^T \bar{U}\right\}.$$

end if

{solve IP master problem}

$min_Y\{\theta \mid$ cuts, $Y \in \{0, 1\}\}$

Lower bound $:= \bar{\theta}$

end while

7.4 A Case Study for Hurricanes

The IEEE 118-bus system is considered to study the proposed model. The component of the system located on the path of the upcoming hurricane, along with the associated probability of damage, as well as the scale parameter of the Weibull distributed time to repair, are given in the first three columns of Tables 7.1–7.3. The shape parameter of the Weibull distribution for all components is assumed to be equal to 1. The value of lost load is $3706/MWh for industrial loads, $6979/MWh for commercial loads, and $110/MWh for residential areas [30]. The load on bus B62 is industrial, while loads on buses B88, B92, and B93 are commercial. The rest of the loads are considered residential. The repair crew is considered to be the only limited resource that is allocated to repair damaged components. It is assumed that each damaged substation requires 10 repair crews/hour, while each damaged transmission line requires 15 repair crews/hour. The wages for repair crews of substations are assumed as follows: $60/hour at shift 1 (8:00 A.M.–4:00 P.M.), $70/hour at shift 2 (4:00 P.M.–12:00 A.M.), and $80/hour at shift 3 (12:00 A.M.–8:00 A.M.). An incremental rate of $5/hour is added to the wage of the transmission lines' repair crew. The generation cost is considered to be $35.09/MWh [31]. The shutdown cost for each generating unit is assumed to be $250. The startup cost is assumed to be $150 within the first hour after the last shutdown. For each additional hour (up to eight hours), an incremental cost of $25 is added to the startup cost. Three cases are considered for analysis as follows:

Table 7.1 Probability of damage and scale parameter of time to repair for generating units.

Unit number	Damage probability	TTR scale parameter
G26	0.15	8
G38	0.35	8
G39	0.50	16
G40	0.45	16
G41	0.35	8
G42	0.20	8
G43	0.80	12
G44	0.25	12
G45	0.40	12

Table 7.2 Probability of damage, time to repair parameter, and the derived resource allocation for buses.

Bus number	Damage probability	TTR scale parameter	Schedule (case I)	Schedule (case II)	Schedule (case III)
*B*62	0.70	10	1–10	1–10	1–8
*B*85	0.20	10	1–9	1–9	1–3
*B*86	0.40	10	1–8	1–8	1–5
*B*87	0.15	7	25–28 46–65	N/A	9–10
*B*88	0.1	10	N/A	N/A	3–4
*B*89	0.05	10	96–107	N/A	1–2
*B*90	0.60	10	1–18	1–18	1–7
*B*91	0.10	7	3–4	5–6	6–7
*B*92	0.30	10	1–4	1–4	1–4
*B*93	0.05	7	1–8	1–8	2–3
*B*94	0.40	10	1–30	1–30	1–5
*B*95	0.20	10	1–8	1–8	2–4
*B*96	0.25	10	1–11	1–11	1–4

Case I (full restoration): The primary resources are considered unconstrained. The restoration of all components is enforced. The aim is to find the optimal value for the maximum resources required for full restoration.

Case II (partial restoration): The primary resources are constrained to obtain value in Case I. The restoration is not enforced for all damaged components. The aim is to analyze the economic dynamics of the restoration, regardless of system-level reliability.

Case III (expected value problem): The expected value problem of the proposed model is solved. The aim is to find the value of a stochastic solution (VSS) for the problem.

All three cases are analyzed in a 120-hour restoration planning horizon. For Cases I and II, 3000 independent scenarios are generated using the Latin hypercube sampling method [32]. With the backward reduction algorithm [28], the number of scenarios is reduced to 10, with associated probabilities of 0.336, 0.011, 0.017, 0.065, 0.029, 0.308, 0.115, 0.012, 0.057, and 0.05. All three cases are solved using the Benders decomposition method. First, the proposed model is solved for Case I to find $R_{(t)}^{max}$, i.e., the maximum resource level required to restore the entire

Table 7.3 Probability of damage, time to repair scale parameter and the derived resource allocation for lines.

Line number	Damage probability	TTR scale parameter	Schedule (case I)	Schedule (case II)	Schedule (case III)
*L*91	0.20	10	1–3 67–90	N/A	3–5
*L*92	0.30	10	41–63	N/A	5–8
*L*100	0.10	10	1–2 25–38	N/A	1–2
*L*101	0.35	10	1–8	N/A	5–9
*L*131	0.70	10	46–87 1–4	N/A	2–9
*L*132	0.55	10	1–8 72–81	1–8	1–7
*L*133	0.30	15	1–3	2–4	1–6
*L*134	0.25	15	5–7	N/A	2–6
*L*135	0.35	10	1–13 95–107	1–13	3–7
*L*136	0.20	10	2–6	N/A	1–3
*L*137	0.20	15	44–74	N/A	1–4
*L*138	0.15	15	70–89	N/A	4–7
*L*139	0.15	15	24–38	N/A	4–7
*L*140	0.35	10	97–106	N/A	1–5

system. After obtaining $R^{max}_{(t)}$, this value is imposed as a constraint in Case II to study system behavior, when the system-level reliability is not considered. The derived value for $R^{max}_{(t)}$ in Case I is 210 crew/hour. The fourth and fifth columns of Table 7.2, respectively, show the optimal schedule of resources that need to be allocated to damaged substations in Cases I and II, while the fourth and fifth columns of Table 7.3, respectively, represent the allocation of resources to transmission lines in Cases I and II. The adjacent allocation schedules to each component are merged, while the overlapping allocations are removed from the results (associated costs are also deducted in the latter case). As shown in the results, there are components in the system with multiple allocations, i.e., B87, L91, L100, L131, L132, and L135 in Case I. These multiple allocations perform as insurance for the system for different scenarios that can occur when a hurricane strikes. On the other hand, there are components in the system without any allocated resources. The reason is the low expected cost of damage to these components. This phenomenon also occurs in Case II due to partial restoration, which ignores the system-level reliability. Due to the economic dynamics of the system, i.e., the cost sensitivity of the system to the functional state of each component, Case II does not allocate resources to some components of the system that do not have considerable expected economic risk. Furthermore, the presence of redundant components in the system that can compensate for the offline state of other components is another reason for the observed behavior. However, from the system-level reliability perspective, the restoration scheme of Case II is not always preferable.

For Case III, rather than deriving a scenario-based solution, the expected value of the parameters is plugged into the proposed model. The last columns of Tables 7.2 and 7.3 show the optimal resource allocation in Case III for substations and lines, respectively. As shown in Figure 7.1, the total expected restoration cost in Case III is $15,581,870, which is higher than in Cases I and II (i.e., $15,343,980 and $15,065,320, respectively). Therefore, the value of the stochastic solution, which is the difference between the stochastic solution and the expected value solution, is $237,890 and $516,550 for Cases I and II, respectively. The expected load interruptions for Cases I, II, and III are 1469, 1467, and 1761 MWh, respectively. While the expected load interruption cost for Case III is significantly higher than for Cases I and II, the generation cost of Case I is slightly lower than two other cases. As shown in Figure 7.2, the optimal resource level for all three cases starts with a high value but dramatically drops by the end of the first working shift. As results show, Cases I and II have a similar pattern for optimal

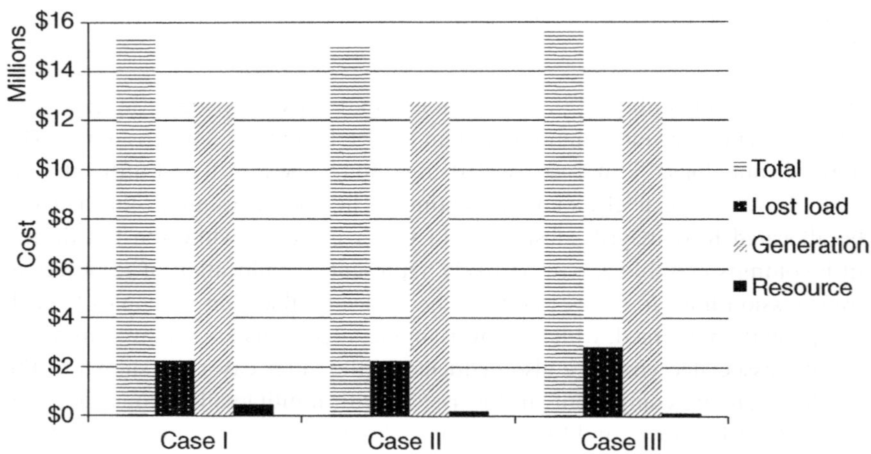

Figure 7.1 Expected cost breakdown for three scenarios.

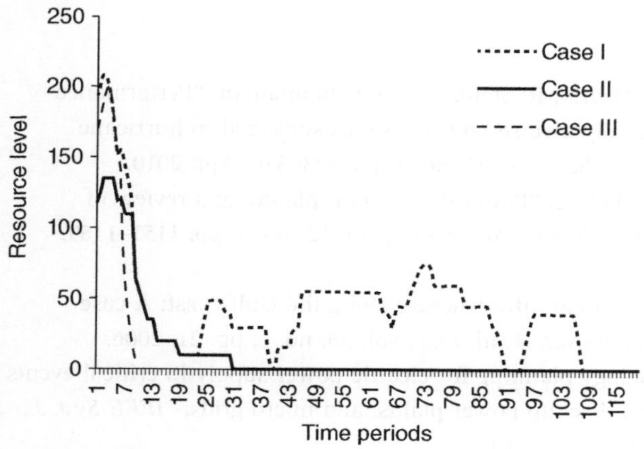

Figure 7.2 Optimal resource level for three scenarios.

resource allocation from the beginning of shift 2 until the end of shift 3. The total resource costs for Cases I, II, and III are $373,575, $114,450, and $70,725, respectively. While Case I has the highest, and Case III has the lowest resource allocation cost, Case II has the most cost-effective strategy to restore the system. However, due to the contingency of the system to unexpected failures and faults, the partial restoration strategy of Case II does not provide the desired system-level reliability in normal operating conditions. On the other hand, the full restoration strategy of Case I provides higher system-level reliability at the expense of a higher resource allocation cost. Considering this trade-off, decision-makers can choose the desirable strategy based on their system operation preferences.

7.5 Summary

A stochastic model was introduced to support the decision-making process for power system restoration in the pre-hurricane phase. The model was formulated as a two-stage stochastic problem with complete recourse. After scenario reduction, the large-scale equivalence of the universe problem was solved using Benders decomposition. Two strategies, i.e., full and partial restoration, were analyzed and the value of the stochastic solution was calculated. The value of stochastic solution as an index justifies the advantage of obtaining stochastic solution over expected value solution. The numerical results demonstrate the merits and disadvantages of each strategy. While the partial restoration strategy provides a more cost-effective restoration plan, it may not provide the same system-level reliability that the full restoration strategy secures. However, decision-makers can choose the best strategy based on the operations policy of the utility company.

References

1 J. Winkler, L. Duenas-Osorio, R. Stein, and D. Subramanian, "Performance assessment of topologically diverse power systems subjected to hurricane events," *Reliab. Eng. Syst. Saf.*, vol. 95, no. 4, pp. 323–336, Apr. 2010.

2 S. K. Khator and L. C. Leung, "Power distribution planning: a review of models and issues," *IEEE Trans. Power Syst.*, vol. 12, no. 3, pp. 1151–1159, Aug. 1997.

3 B. Ball, "Rebuilding electrical infrastructure along the Gulf coast: A case study," *Bridge-Washington-Nat. Acad. Eng.*, vol. 36, no. 1, pp. 21, 2006.

4 A. Kwasinski, "Technology planning for electric power supply in critical events considering a bulk grid, backup power plants, and micro-grids," *IEEE Syst. J.*, vol. 4, no. 2, pp. 167–178, Jun. 2010.

5 N. Xu, S. D. Guikema, R. A. Davidson, L. K. Nozick, Z. Cagnan, and K. Vaziri, "Optimizing scheduling of post-earthquake electric power restoration tasks," *Earthquake Eng. Struct. Dyn.*, vol. 36, no. 2, pp. 265–284, Feb. 2007.

6 N. Perrier *et al.*, "A survey of models and algorithms for emergency response logistics in electric distribution systems. Part I: Reliability planning with fault considerations," *Comput. Oper. Res.*, vol. 40, no. 7, pp. 1895–1906, Jul. 2013.

7 N. Perrier *et al.*, "A survey of models and algorithms for emergency response logistics in electric distribution systems. Part II: Contingency planning level," *Comput. Oper. Res.*, vol. 40, no. 7, pp. 1907–1922, Jul. 2013.

8 P. J. Maliszewski and C. Perrings, "Factors in the resilience of electrical power distribution infrastructures," *Appl. Geogr.*, vol. 32, no. 2, pp. 668–679, Mar. 2012.

9 S. M. Quiring, L. Zhu, and S. D. Guikema, "Importance of soil and elevation characteristics for modeling hurricane-induced power outages," *Nat. Hazards*, vol. 58, no. 1, pp. 365–390, Jul. 2011.

10 A. R. Dorothy , K. C. Kapur, and R. D. Christie, "Methodology for assessing the resilience of networked Infrastructure," *IEEE Syst. J.*, vol. 3, no. 2, pp. 174–180, Jun. 2009.

11 S. Bjarnadottir, Y. Li, and M.G. Stewart, "Hurricane risk assessment of power distribution poles considering impacts of a changing climate," *J. Infrastruct. Syst.*, vol. 19, no. 1, pp. 12–24, Mar. 2012.

12 H. Liu, R. A. Davidson, and T. Apanasovich, "Statistical forecasting of electric power restoration times in hurricanes and ice storms," *IEEE Trans. Power Syst.*, vol. 22, no. 4, pp. 2270–2279, Nov. 2007.

13 H. Liu, R. Davidson, D. Rosowsky, and J. Stedinger, "Negative binomial regression of electric power outages in hurricanes," *J. Infrastruct. Syst.*, vol. 11, no. 4, pp. 258–267, Dec. 2005.

14 S. D. Guikema, S. M. Quiring, and S. R. Han, "Prestorm estimation of hurricane damage to electric power distribution systems," *Risk Anal.*, vol. 30, no. 12, pp. 1744–1752, Oct. 2010.

15 M. J. Yao and K. J. Min, "Repair-unit location models for power failures," *IEEE Trans. Eng. Manag.*, vol. 45, no. 1, pp. 57–65, Feb. 1998.

16 C. Coffrin, P. V. Hentenryck, and R. Bent, "Strategic stockpiling of power system supplies for disaster recovery," in *Proc. IEEE Power Eng. Soc. Gen. Meeting*, San Diego, CA, USA, Jul. 2011.

17 Y. Yongbo *et al.*, "Service restoration with consideration of rush repair," in *Proc. Power Eng. Auto. Conf. (PEAM)*, Wuhan, China, Sep. 2011.

18 S. Wang, B. R. Sarker, L. Mann Jr, and E. Triantaphyllou, "Resource planning and a depot location model for electric power restoration," *Eur. J. Oper. Res.*, vol. 155, no. 1, pp. 22–43, May 2004.

19 D. Lubkeman and D. E. Julian, "Large scale storm outage management," in *Proc. IEEE Power Eng. Soc. Gen. Meeting*, Denver, CO, USA, Jun. 2004.

20 X. Wang *et al.*, "Pre-disaster prevention strategies for coupled thermoelectric systems under extreme cold disasters," in *2022 Power Syst. Green Energy Conf. (PSGEC)*, Shanghai, China, 2022, pp. 836–842, doi: https://doi.org/10.1109/PSGEC54663.2022.9881150.

21 J. Su, R. Zhang, P. Dehghanian, and M. H. Kapourchali, "Pre-disaster allocation of mobile renewable-powered resilience-delivery sources in power distribution networks," in *2023 North American Power Symp. (NAPS)*, Asheville, NC, USA, 2023, pp. 1–6, doi: https://doi.org/10.1109/NAPS58826.2023.10318581.

22 H. Hou *et al.*, "Stochastic pre-disaster planning and post-disaster restoration to enhance distribution system resilience during typhoons," *Energy Convers. Econ.*, vol. 4, no. 5, pp. 346–363, 2023.

23 P. Kall, and J. Mayer, *Stochastic Linear Programming: Models, Theory, and Computation*. New York, NY, USA: Springer, 2011.

24 G. Laporte and F. V. Louveaux, "The integer L-shaped method for stochastic integer programs with complete recourse," *Oper. Res. Lett.*, vol. 13, no. 3, pp. 133–142, Apr. 1993.

25 M. Carrion and J. M. Arroyo, "A computationally efficient mixed-integer linear formulation for the thermal unit commitment problem," *IEEE Trans. Power Syst.*, vol. 21, no. 3, pp. 1371–1378, Aug. 2006.

26 J. R. Birge and F. V. Louveaux, *Introduction to Stochastic Programming*. New York, NY, USA: Springer, 1997.

27 M. D. McKay, R. J. Beckman, and W. J. Conover, "Comparison of three methods for selecting values of input variables in the analysis of output from a computer code," *Technometrics*, vol. 21, no. 2, pp. 239–245, 1979.

28 H. Heitsch and W. Römisch “Scenario reduction algorithms in stochastic programming,” *Comput. Optim. Appl.*, vol. 24, no. 2–3, pp. 187–220, 2003.

29 E. Kalvelagen, “Benders decomposition with GAMS,” [Online]. Available: http://www.amsterdamoptimization.com/pdf/benders.pdf

30 London Economics International LLC, “Estimating the value of lost load: Briefing paper prepared for the ERCOT by LEI,” Boston, MA, USA, 2013.

31 U.S. Energy Information Administration, 2013. Accessed: Sep 2019. [Online]. Available: http://www.eia.gov

32 G. D. Wyss and K. H. Jorgensen, “A user’s guide to LHS: Sandia’s Latin hypercube sampling software,” Sandia National Lab., Albuquerque, NM, USA, 1998.

8

The Economics of Contingency Planning Under Extreme Events

8.1 Background

Climate shocks and natural disasters can cause temporary disruption or permanent damage to power generation infrastructure. One of the approaches for mitigating the risk of these events, for example, to prevent subsequent transmission flow violations, is to carry out topology control (TC). Preventive and corrective TC can provide economic and reliability benefits compared to other control methods, such as generation unit rescheduling or load shedding [1, 2]. In [3], TC was considered by applying current sources at certain bus terminals in the base network. The injected currents were represented as control variables in the optimal TC problem. In [4], TC was employed as a corrective action to mitigate contingencies. Also, TC was used to model outages in optimal power flow. In [5], a TC method was devised for calculating $N-1$ secure states. A linear switching model was applied to model control actions applied to contingency constraints. In [6], TC was considered as a means of mitigating violations in transmission flows and bus voltages. Also, practical issues related to switching operations were addressed. In these papers, transformer tap adjustments and static VAR compensators were recognized as a means of controlling voltage problems [6]. In [7], a TC algorithm was developed for switching off lines and buses to mitigate contingencies. This TC algorithm was based on a sparse inverse technique and fast decoupled power flow. Also, in [8], TC provided system operators with a congestion management tool.

These works showed that TC provides flexible control actions for voltage stability, congestion management, loss reduction, and system security. However, TC can provide economic benefits in a market context [9]. In [10] and [11], the problem of finding an optimal generation dispatch and transmission topology was investigated. In [10], the TC problem was solved only for the base case dispatch of the system, while in [11], the $N-1$ contingency criterion was also considered. A mixed-integer programming (MIP) model was used, which employed binary variables to represent the transmission line state. These papers found that TC could achieve large improvements in dispatch costs. The work in [12] extended [10], demonstrating that market participants are subject to system uncertainties when considering TC. The paper discussed how topology changes could affect nodal prices, load payments, generation revenues, congestion costs, and flowgate prices.

Authors in [13] presented a probabilistic framework to improve power system TC by using transmission line switching. The goal is to optimize economic efficiency and reliability, especially considering the uncertainties of renewable energy generation and load variability. By employing the point estimation method in a DC optimal power flow model, the approach considers uncertainties to provide more realistic switching solutions compared to deterministic methods. The framework includes a decision-making tool to determine the optimal number of switching actions per hour, balancing economic gains against reliability risks. In [14], a linearized AC optimal power flow model to tackle the challenges of stochastic security-constrained unit commitment (SCUC) was proposed. The model includes transmission switching to further increase grid flexibility and reduce uncertainty in wind power generation. The proposed model for TC aims to alleviate transmission congestion, thereby reducing total operating costs, average locational marginal prices, and energy-weighted equivalent system cost, while improving wind power utilization and enhancing power system reliability by decreasing Expected Energy Not Served (EENS). Ref. [15] investigated the use of the Dueling Deep Q-Networks (DDQN) reinforcement learning algorithm for controlling power systems through line switching and substation TC to maintain stable line flow under various scenarios. Authors highlighted the advantage of DDQN over traditional Deep Q-Networks resulting from learning nonessential states, enabling faster convergence and reduced training time. Ref. [16] introduced a methodology for improving the power system resilience by reconfiguring the system via optimally switching transmission lines based on optimal DC power flow. The proposed methodology quantifies the contingency index to identify system vulnerabilities and prioritize contingencies by impact. The proposed model aims at helping system operators to identify critical elements requiring attention for improved resilience and informs future system expansion plans. While the current method addresses several contingencies, it suggests further research to handle scenarios involving island formation or generator disconnection.

This chapter considers the optimal TC for alleviating overloads in SCUC, which would consider temporal generating unit and transmission network constraints. Figure 8.1 depicts the hierarchy for calculating TC-enabled SCUC. The large-scale SCUC solution can represent a computation burden in power system operations. Since it is generally considered impractical to solve the generation unit and the entire set of network constraints together in SCUC, various decomposition approaches are considered by commercial SCUC packages.

The proposed decomposition applied to the SCUC problem in this chapter consists of the UC problem and the TC problem, which are decomposed using a Benders decomposition method. The optimality of Benders decomposition, as well as its applicability to power system problems in practical cases, are discussed in [17]–[21].

The UC problem in Figure 8.1 will find the optimal schedule of units, considering the prevailing UC constraints. The initial optimal schedule of generating units is obtained based on the available market data. The UC solution is used in the TC problem to find the optimal hourly dispatch of units, considering transmission constraints and the switching capability of transmission lines.

The TC problem in Figure 8.1 consists of three main blocks, as shown in Figure 8.2. The TC Feasibility Check inspects the UC result to find whether a feasible TS solution can be found in the base case without considering contingencies. Benders cuts are generated and added to the UC problem if violations persist. After satisfying the TC feasibility, the Optimal TC Scheduling block will utilize the UC solution to find the optimal dispatch of generating units and the state of switchable lines in the base case and contingencies. The Transmission Contingencies Check block will then examine the Optimal TC Scheduling results to determine whether a converged DC power flow solution can be obtained. The Transmission Contingencies Check block will also adjust the transmission flows to mitigate any bus mismatches. In the case of violations, Benders cuts are formed and added to the Optimal TC Scheduling block for the redispatch of generating units.

If a converged DC power flow solution for a contingency considered in the TC problem can be found, this contingency will be labeled as controllable. However, if a converged DC power flow solution does not exist after a certain number of iterations between the Optimal TC Scheduling and the Transmission Contingencies Check blocks, i.e., the violations persist in the Transmission Contingencies Check block, the considered contingency will be labeled as uncontrollable. The controllable contingencies would not require further processing as the existing

Figure 8.1 SCUC using TC.

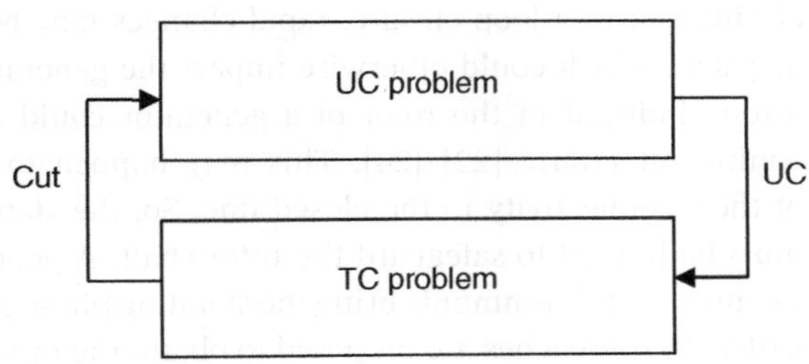

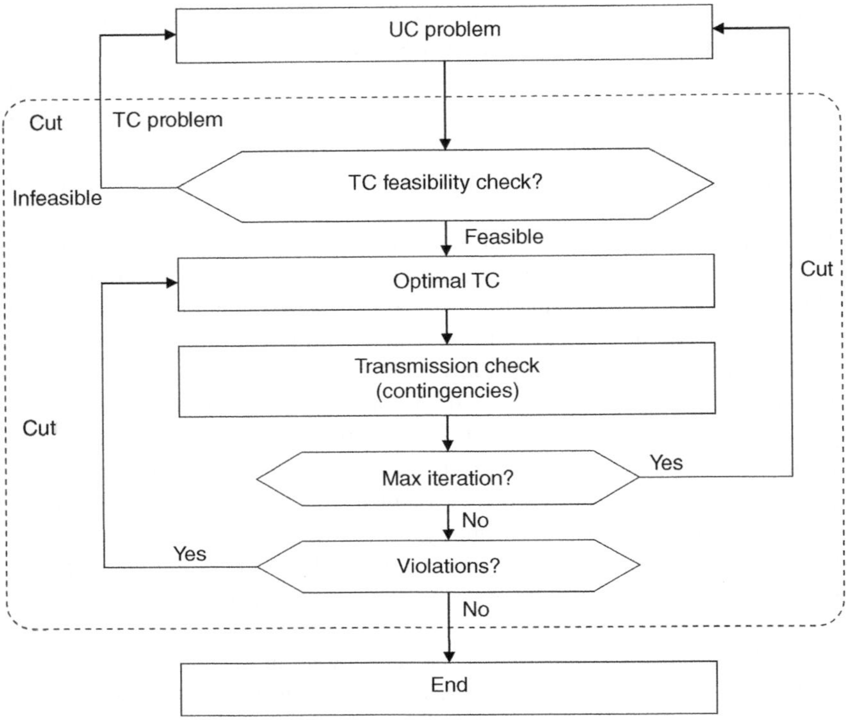

Figure 8.2 Flowchart of SCUC using TC.

UC solution can handle them. The uncontrollable contingencies, which cannot be handled in the TC problem, are sent back to the UC problem to find a preventive UC schedule. Labeling the contingencies, the controllable ones are considered first (corrective), and then uncontrollable ones are handled (preventive). In other words, the proposed approach examines less expensive corrective actions before resorting to the more expensive preventive actions.

One of the concerns in performing the TC in successive hours is the possibility of excessive standing phase angle difference across a switched line. If a line closing causes a loop closure, rapid changes may be required in the generation dispatch, which could otherwise impact the generator shaft. The corresponding torque induced in the rotor of a generator could cause generator fatigue and equipment failure [22]–[24]. This may happen to any generators irrespective of their connectivity to the closed line. So, the standing phase angle difference must be limited to safeguard the rotor shaft. A generation redispatch upon line closures would commonly bring the standing phase angles back to the safe region. Different approaches are proposed to obtain the minimum generation redispatch needed to achieve desired standing phase angles in a reasonable time [25]–[29]. This problem, which is inevitable in restoration practices, may also occur in the normal operation of power systems when attempting to reclose a single line that is part of a transmission loop [24]. To present a practical TC model, the standing phase angle difference limit of switchable lines is formulated in the proposed model.

8.2 An Economic Model for Contingency Planning

The SCUC problem depicted in Figure 8.2 is formulated as follows:

8.2.1 UC Problem (Optimal Hourly Schedule of Units)

The objective of the UC problem is to determine the day-ahead schedule of generating units to minimize the system operating cost while satisfying the prevailing constraints. A complete list of UC constraints and optimization techniques is found in [13].

8.2.2 TC Problem

The TC problem consists of three parts: TC Feasibility Check, Optimal TC Scheduling, and Transmission Check, as discussed below.

8.2.2.1 TC Feasibility Check

TC Feasibility Check, which is an MIP problem, examines the possibility of finding a feasible TC solution. The objective is defined as

$$Min \sum_{t=1}^{NT} \sum_{b=1}^{NB} (SL_{bt,1} + SL_{bt,2}) \tag{8.1}$$

where

t and b, respectively denote time and bus indices
NT and NB, respectively represent the number of time periods and buses

The bus power mismatch is presented by (8.2), where $SL_{bt,1}$ and $SL_{bt,2}$ are surplus and deficit, s is the index for switchable lines, and ns is the index for non-switchable lines. Also, P_{it} represents the generation of unit i at time t, PL^{s}_{lt} denotes the power flow of switchable line l at time t, PL^{ns}_{lt} represents the power flow of non-switchable line l at time t, and Pd_{bt} signifies the load at bus b at time t. NS represents the number of switchable lines, while NNS denotes the number of non-switchable lines. The other constraints are the power flow of switchable lines (8.3)–(8.6) where δ_{mt} is the phase angle of bus m at time t, x_l is the reactance of line l, z_{lt} is the switching state of line l at time t, NL is the number of lines and $PL_{l,\max}$ is the maximum capacity of line l. The constraints for the power flow of non-switchable lines are presented in (8.7)–(8.9). The standing phase angle difference limits are (8.10)–(8.11), where Δ_l is the maximum standing phase angle difference of line l and M is a large positive number. The non-islanding constraint is represented in (8.12).

$$\sum_{i\in U_b} P_{it} - \sum_{l\in L_b} PL^s_{lt} - \sum_{l\in L_b} PL^{ns}_{lt} - Pd_{bt} + SL_{bt,1} - SL_{bt,2} = 0$$
$$(b = 1, \dots, NB)(s = 1, \dots, NS)(ns = 1, \dots, NNS) \tag{8.2}$$

$$PL^s_{lt} - (\delta_{mt} - \delta_{nt})/x_l + M(1 - z_{lt}) \geq 0 \quad (t = 1, \dots, NT)(l = 1, \dots, NL)$$
$$(s = 1, \dots, NS) \tag{8.3}$$

$$PL^s_{lt} - (\delta_{mt} - \delta_{nt})/x_l - M(1 - z_{lt}) \leq 0 \quad (t = 1, \dots, NT)(l = 1, \dots, NL)$$
$$(s = 1, \dots, NS) \tag{8.4}$$

$$PL^s_{lt} \leq PL_{l,max} z_{lt} \quad (t = 1, \dots, NT)(l = 1, \dots, NL)(s = 1, \dots, NS) \tag{8.5}$$

$$-PL^s_{lt} \leq PL_{l,max} z_{lt} \quad (t = 1, \dots, NT)(l = 1, \dots, NL)(s = 1, \dots, NS) \tag{8.6}$$

$$PL^{ns}_{lt} = (\delta_{mt} - \delta_{nt})/x_l \quad (t = 1, \dots, NT)(l = 1, \dots, NL)(ns = 1, \dots, NNS) \tag{8.7}$$

$$PL^{ns}_{lt} \leq PL_{l,max} \quad (t = 1, \dots, NT)(l = 1, \dots, NL)(ns = 1, \dots, NNS) \tag{8.8}$$

$$-PL^{ns}_{lt} \leq PL_{l,max} \quad (t = 1, \dots, NT)(l = 1, \dots, NL)(ns = 1, \dots, NNS) \tag{8.9}$$

$$\delta_{mt} - \delta_{nt} \leq \Delta_l + Mz_{l(t-1)} + M(1 - z_{lt}) \quad (t = 1, \dots, NT)(l = 1, \dots, NL) \tag{8.10}$$

$$\delta_{mt} - \delta_{nt} \geq -\Delta_l - Mz_{lt} - M(1 - z_{l(t-1)}) \quad (t = 1, \dots, NT)(l = 1, \dots, NL) \tag{8.11}$$

$$\sum_{l\in L_b} z_{lt} \geq 1 \quad (t = 1, \dots, NT) \tag{8.12}$$

The TC constraints use a single binary variable z_{lt}. When this variable equals one, power flow constraints on switchable lines will be the same as those of other lines. In such a case, the line will be treated the same as other lines. When the line's binary variable is zero, Eqs. (8.3)–(8.6) would impose a zero line flow and switch the line off. The direction of transmission flow is from bus m to bus n. Constraints (8.10) and (8.11) consider the preset limits for the standing phase angle difference. The standing phase angle difference must be within its limits before an attempt is made to close breakers. Using Eqs. (8.10) and (8.11), standing phase angle difference limits are imposed when a line is switched on. Using Eq. (8.12), the islanding is prevented. This constraint guarantees that at least one line is connected to each bus when all corresponding lines are switchable.

If the TC Feasibility Check is infeasible, a linear programming problem is formed by setting the state of switchable lines to 1 and considering transmission constraints (8.2)–(8.9). This problem minimizes the bus power mismatches for 24 hours and forms Benders cuts for the UC problem. This problem applies the UC schedule using Eq. (8.13).

$$P_{it} = \widehat{P}_{it}\widehat{I}_{it} \leftrightarrow \pi_{it} \quad (i = 1, \ldots, NG) \tag{8.13}$$

where $\widehat{P}_{it}$ and $\widehat{I}_{it}$ are fixed values calculated by the UC problem.

In Eq. (8.13), the arrow refers to the duality of the equality constraint, where π_{it} is the dual variable (also known as simplex multiplier) of equality constraint (8.13) and I_{it} is the commitment state of unit i at time t. Mathematically, this multiplier represents the marginal increment/decrement of the objective value when $\widehat{P}_{it}\widehat{I}_{it}$ is changed. This dual variable is needed to form the Benders cut. The solution to this problem will provide the hourly cuts for the UC problem stated as

$$\widehat{v}_t + \sum_{i=1}^{NG} \pi_{it}(P_{it}I_{it} - \widehat{P}_{it}\widehat{I}_{it}) \leq 0 \tag{8.14}$$

In Eq. (8.14), the hourly bus power mismatch is noted asv_t and NG is the number of units. The cut represents the coupled information on the unit generation dispatch and commitment state. The current violations can be mitigated by recalculating the hourly commitment states and redispatching the generating units.

If the TC Feasibility Check solution is feasible, we will proceed to the Optimal TC Scheduling block as discussed next.

8.2.2.2 Optimal TC Scheduling

The Optimal TC Scheduling problem will calculate the optimal dispatch of generating units and the switchable line states in the base case and contingencies, given the hourly UC schedule

$$Min \sum_{i=1}^{NG} \sum_{t=1}^{NT} \left[F_{ci}\left(P_{it}^{c}\right)\widehat{I}_{it}\right] \tag{8.15}$$

which is subject to unit and system constraints in which F_{ci} is the production cost function of unit i for contingency c. The unit constraints include generation bids, power generation, and ramp-up/down-rate limits. System constraints include power balance and spinning/operating reserve requirements, as well as fuel and emission constraints. A complete list of constraints can be found in [17]. The objective function (8.15) considers fixed unit commitment states, which are calculated in the UC problem. In the case of contingencies, the objective is also subject to load balance (8.16), where D_t is the system demand at time t, and generation limit (8.17) constraints where $P_{i,\,min}$ and $P_{i,\,max}$ are the minimum and maximum power generation of unit I, respectively.

$$\sum_{i=1}^{NG} P_{it}^{c} = D_t \quad (t = 1, \ldots, NT) \tag{8.16}$$

$$P_{i,min} I_{it} UX_{it}^{c} \leq P_{it}^{c} \leq P_{i,max} I_{it} UX_{it}^{c} \quad (t = 1, \ldots, NT)(i = 1, \ldots, NG) \tag{8.17}$$

where UX_{it}^{c} is the contingency state of unit i at time t in contingency c. Note that $c = 0$ represents the base case. So, we would have $UX_{it}^{0} = 1$. In Eq. (8.16), the load balance equation is satisfied for each contingency. According to Eq. (8.17), when a unit is on outage, i.e., a contingency, its generation would be zero.

The Optimal TC Scheduling is a MIP problem because of the binary states of switchable lines. At the first iteration, there are no constraints on line flows or switchable line states. So, any values could be assigned to these variables. But in the subsequent iterations, the Benders cuts from the Transmission Contingencies Check establish constraints on switchable line states.

8.2.2.3 Transmission Check in Contingencies

The solution of the Optimal TC Scheduling block will be checked in the Transmission Contingencies Check to see whether a converged DC power flow solution can be obtained. So, we would have

$$Min\ w_t^c = \sum_{b=1}^{NB} \left(SL_{bt,1}^{c} + SL_{bt,2}^{c} \right) \tag{8.18}$$

In Eq. (8.18), w_t^c is the power mismatch at time t and contingency c. This objective is subject to transmission security constraints (8.19)–(8.32).

$$P_{it}^{c} = \widehat{P}_{it}^{c} \leftrightarrow \pi_{it}^{c} \quad (i = 1, \ldots, NG) \tag{8.19}$$

$$z_{lt}^{c} = \widehat{z}_{lt}^{c} \leftrightarrow \mu_{lt}^{c} \quad (l = 1, \ldots, NL) \tag{8.20}$$

$$\sum_{i \in U_b} P_{it}^{c} - \sum_{l \in L_b} PL_{lt}^{sc} - \sum_{l \in L_b} PL_{lt}^{nsc} - Pd_{bt} + SL_{bt,1}^{c} - SL_{bt,2}^{c} = 0$$
$$(b = 1, \ldots, NB)(s = 1, \ldots, NS)(ns = 1, \ldots, NNS) \tag{8.21}$$

$$PL_{lt}^{sc} - \left(\delta_{mt}^{c} - \delta_{nt}^{c}\right)/x_l + M\left(1 - z_{lt}^{c}\right) + M\left(1 - UY_{lt}^{c}\right) \geq 0$$
$$(l = 1, \ldots, NL)(s = 1, \ldots, NS) \tag{8.22}$$

$$PL_{lt}^{sc} - \left(\delta_{mt}^{c} - \delta_{nt}^{c}\right)/x_l - M\left(1 - z_{lt}^{c}\right) - M\left(1 - UY_{lt}^{c}\right) \leq 0$$
$$(l = 1, \ldots, NL)(s = 1, \ldots, NS) \tag{8.23}$$

$$PL_{lt}^{sc} \leq PL_{l,max} z_{lt}^{c} UY_{lt}^{c} \quad (l = 1, \ldots, NL)(s = 1, \ldots, NS) \tag{8.24}$$

$$-PL_{lt}^{sc} \leq PL_{l,max} z_{lt}^{c} UY_{lt}^{c} \quad (l = 1, \ldots, NL)(s = 1, \ldots, NS) \tag{8.25}$$

$$PL_{lt}^{nsc} - \left(\delta_{mt}^{c} - \delta_{nt}^{c}\right)/x_l + M\left(1 - UY_{lt}^{c}\right) \geq 0$$
$$(l = 1, \ldots, NL)(s = 1, \ldots, NS) \tag{8.26}$$

$$PL_{lt}^{nsc} - \left(\delta_{mt}^{c} - \delta_{nt}^{c}\right)/x_l - M\left(1 - UY_{lt}^{c}\right) \leq 0$$
$$(l = 1, \ldots, NL)(s = 1, \ldots, NS) \tag{8.27}$$

$$PL_{lt}^{nsc} \leq PL_{l,max} UY_{lt}^{c} \quad (l = 1, \ldots, NL)(ns = 1, \ldots, NNS) \tag{8.28}$$

$$-PL_{lt}^{nsc} \leq PL_{l,max} UY_{lt}^{c} \quad (l = 1, \ldots, NL)(ns = 1, \ldots, NNS) \tag{8.29}$$

$$\delta_{mt}^{c} - \delta_{nt}^{c} \leq \Delta_l + M z_{l(t-1)}^{c} + M\left(1 - z_{lt}^{c}\right) \quad (l = 1, \ldots, NL) \tag{8.30}$$

$$\delta_{mt}^{c} - \delta_{nt}^{c} \geq -\Delta_l - M z_{l(t-1)}^{c} - M\left(1 - z_{lt}^{c}\right) \quad (l = 1, \ldots, NL) \tag{8.31}$$

$$\sum_{l \in L_b} z_{lt}^{c} \geq 1 \quad (t = 1, \ldots, NT) \tag{8.32}$$

where $\widehat{P}_{it}^{c}$ and $\widehat{z}_{lt}^{c}$ are fixed values calculated by Optimal TC Scheduling. Power flows are obtained by (8.22)–(8.25). When a line is switched off, the line flow will be set to zero and removed from power flow equations. The power flow of non-switchable lines is obtained by (8.26)–(8.29). In (8.22)–(8.29), the contingency state of line l at time t is considered by UY_{lt}^{c}. Note that $c = 0$ represents the base case. Therefore, we would have $UY_{lt}^{0} = 1$. Standing phase angle difference limits are considered by (8.30) and (8.31), and islanding is prevented using (8.32) for buses in which all corresponding lines are switchable. Here, (8.30) and (8.31) use the given values of switchable states. So, there is no time-based coupling in this problem.

If the total mismatch is zero, the proposed generation dispatch and switchable line schedule provide a feasible power flow solution that satisfies transmission security constraints. Otherwise, the Benders cut (8.33) will be added to the Optimal TC Scheduling block to mitigate the violations in the next iteration.

$$\widehat{w}_t^{c} + \sum_{i=1}^{NG} \pi_{it}^{c}\left(P_{it}^{c} - \widehat{P}_{it}^{c}\right) + \sum_{l=1}^{NL} \mu_{lt}^{c}\left(z_{lt}^{c} - \widehat{z}_{lt}^{c}\right) \leq 0 \tag{8.33}$$

This cut represents the coupled information on the unit generation dispatch and switchable line schedule. If the current UC solution cannot mitigate the contingency c violations when the maximum number of iterations has been reached, this contingency will be labeled as uncontrollable, and the Benders cut (8.34) will be returned to the UC problem for calculating a preventive generation schedule

$$\widehat{w}_t^{c} + \sum_{i=1}^{NG} \pi_{it}^{c}\left(P_{it} I_{it} - \widehat{P}_{it}^{c} \widehat{I}_{it}\right) \leq 0 \tag{8.34}$$

The controllable contingencies are handled by corrective actions in the Transmission Contingencies Check without requiring revisions to the existing UC solution. The proposed solution procedure is given as follows:

Step 1. Solve the UC problem.
Step 2. Given the UC schedule, check the TC feasibility. If the TC is feasible, then proceed to Step 4.
Step 3. Minimize bus power mismatches if the TC Feasibility Check is infeasible. Form the Benders cuts and go to Step 1.
Step 4. Use the Optimal TC Feasibility Scheduling to calculate the generation dispatch and the switchable line states in the base case and contingencies.
Step 5. Use the Transmission Check block to minimize bus power mismatches by utilizing the Optimal TC Scheduling results. If the mismatch is not zero, add the Benders cuts to the Optimal TC Scheduling for the next iteration. The Transmission Check block handles the controllable contingencies. However, the Benders cut for uncontrollable contingencies is added to Step 1 for the next UC iteration. Stop the process if the total mismatch is zero.

8.3 Case Studies

Two case studies for the six-bus system and the IEEE 118-bus system are analyzed to illustrate the performance of the proposed method. The proposed method was implemented on a 2.4-GHz personal computer using CPLEX 11.0 [30].

8.3.1 Six-Bus System

The six-bus system is shown in Figure 8.3. The objective is to calculate the least cost of dispatch with an hourly fixed load. The characteristics of generators, lines, and the hourly load distribution over the 24-hour horizon are given in Tables 8.1–8.3, respectively. The load is distributed as 20%, 40%, and 40% among buses 3, 4, and 5,

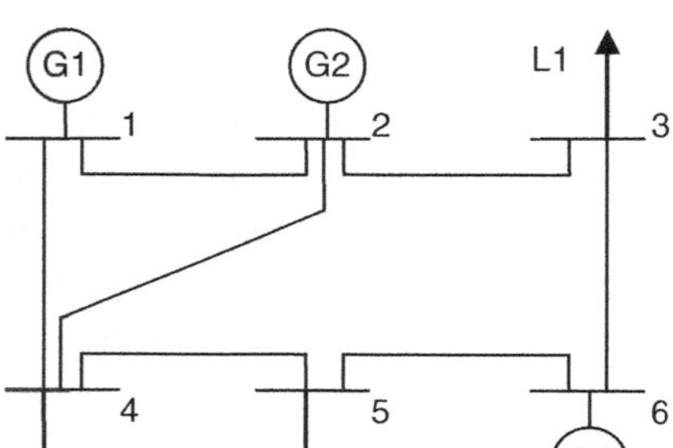

Figure 8.3 Six-bus system.

Table 8.1 Characteristics of generating units.

Unit no.		1	2	3
Bus no.		1	2	6
Cost coefficients	c ($/MW^{2}h)	0.014	0.020	0.086
	b ($/MWh)	19.96	23	29.14
	a ($)	200	150	50
Minimum capacity (MW)		100	10	10
Maximum capacity (MW)		220	200	50
Startup cost ($)		50	40	0
Shutdown cost($)		100	200	0
Minimum up time (h)		4	3	1
Minimum down time (h)		4	2	1
Ramp up rate (MW/h)		40	30	20
Ramp down rate (MW/h)		50	35	20
MSR (MW/min)		2	1.5	0.5
QSC (MW)		15	10	10
Initial hour		+4	+3	+1
Initial generation (MW)		140	20	10

Table 8.2 Characteristics of transmission lines.

Line no.	From bus	To bus	*X* (p.u.)	Flow limit (MW)
1	1	2	0.170	140
2	1	4	0.258	110
3	2	3	0.037	150
4	2	4	0.197	140
5	3	6	0.018	130
6	4	5	0.037	50
7	5	6	0.140	140

Table 8.3 Hourly load demand.

Hour	Load (MW)	Spinning reserve (MW)	Operating reserve (MW)
1	175.19	2.63	12.26
2	165.15	2.48	11.56
3	158.67	2.38	11.10
4	154.73	2.32	10.83
5	155.06	2.33	10.85
6	160.48	2.40	11.23
7	173.39	2.60	12.14
8	177.60	2.85	13.33
9	186.81	3.09	14.39
10	206.96	3.26	15.20
11	228.61	3.43	16.00
12	236.10	3.54	16.52
13	242.18	3.63	16.95
14	243.60	3.66	17.05
15	248.86	3.73	17.42
16	255.79	3.84	17.91
17	256.00	3.84	17.92
18	246.74	3.70	17.27
19	245.97	3.69	17.22
20	237.35	3.56	16.62
21	237.31	3.56	16.62
22	232.67	3.41	15.90
23	195.93	3.02	14.07
24	195.60	2.95	13.78

respectively. In Table 8.1, the maximum sustained rate (MSR) and the quick start capability (QSC) of units are used to satisfy the reserve requirements of the system. The MSR and QSC are used to limit the spinning and operating reserves of the unit, respectively. The 10-minute spinning reserve of a unit is the unloaded synchronized generation that can ramp up in 10 minutes. The spinning reserve of a unit cannot exceed the difference between its maximum capacity and current generation dispatch. This reserve is limited by the 10-minute MSR. Operating reserve is the unloaded synchronized/unsynchronized generating capacity that can ramp up in 10 minutes. When a unit is on, its operating reserve is the same as the spinning reserve. When a unit is off, its operating reserve equals its QSC. The following TC cases are considered:

Case 1: Base case UC (without contingencies)
Case 2: Outage of lines 2–4 is considered in Case 1
Case 3: Outage of unit 2 is considered in Case 1
Case 4: Standing phase angle difference limits are considered in Case 1

The range of standing phase angle differences that a system can withstand mostly depends on the voltage level and is usually determined by steady-state and dynamic simulations. However, in Cases 1, 2, and 3, the maximum standing phase angle is assumed to be large enough to satisfy the associated constraints. In Case 4, the impact of the standing phase angle difference on the UC solution is examined.

Case 1: The UC with a subsequent DC network security check is used to find the results shown in Table 8.4. The cheaper unit 1 is on at all hours, while unit 2 is used at peak hours to satisfy the remaining load and minimize the operating cost. Unit 2 is committed at hour 11 due to a load increase. The transmission network encounters flow violations on lines 1–4 and 4–5. Therefore, expensive unit 3 is turned on to help mitigate violations. Line 4–5 is congested at peak hours 15–19, which leads to a lower dispatch of unit 1 and a higher operating cost. The total operating cost is $125,465.

The TC application will result in a similar schedule, but unit 3 is not committed to the entire scheduling horizon. This UC schedule will satisfy the TC feasibility check. So, the model would proceed to the Optimal TC Scheduling block. Imposing fewer cuts in this case would result in a cheaper UC solution. Using TC, the total operating cost has slightly dropped to $125,362. At hour 11, when lines 1–4 and 4–5 are congested, lines 2–4 and 4–5 are switched off instead of turning on unit 3. There will be no transmission loop in the system, and the existing lines could increase the flows to their limits. Table 8.5 shows the state of switchable lines in which lines 2–4 and 4–5 are mostly switched off. At peak hours 16–17, the line

Table 8.4 UC schedule of six-bus system in case 1.

Units	Hours (0–24)																						
1	1	1	1	1	1	1	1	1	1	1	1	1	1	1	1	1	1	1	1	1	1	1	1	1
2	0	0	0	0	0	0	0	0	0	0	1	1	1	1	1	1	1	1	1	1	1	1	1	0
3	0	0	0	0	0	0	0	0	0	0	1	0	0	0	0	0	0	0	0	0	0	0	0	0

Table 8.5 Line schedule of six-bus system in case 1 using TC.

Line	Hours (0–24)																								
2–4	—	1	1	1	1	1	1	0	0	0	0	0	0	0	0	0	1	1	0	0	0	0	0	0	0
4–5	—	1	1	1	1	1	1	0	0	0	0	0	0	0	0	0	1	1	0	0	0	0	0	0	0

flow constraints cannot be satisfied without replacing the lines 2–4 and 4–5. The standing phase angle differences for the two lines are 10.24° and 1.85°, respectively.

Case 2: In this case, the outage of line 2–4 is considered. The UC in Case 1 cannot satisfy the DC network security check. The new UC is mostly the same as the schedule in Case 1, but unit 2 is committed at hours 10 and 24. This UC will mitigate the flow violations and satisfy the transmission security with a total operating cost of $125,848. The UC solution given in Case 1 will satisfy the optimal TC scheduling block with line switching. The unit schedule is the same as that of Case 1. The TC schedule is different from that of Case 1, where line 4–5 is in service at peak hours 16–17 and off otherwise. The standing phase angle difference of line 4–5 at hour 16 is 0.3°. The line 2–4 contingency is handled with a corrective action at the Transmission Contingencies Check block. The congestion of line 1–2 is mitigated at peak hours 16–17 with a total operating cost of $125,470 (i.e., 0.3% improvement).

Case 3: When unit 2 is on outage, unit 1 will not be able to satisfy the hourly load. Therefore, unit 3 is committed additionally. The preventive schedule without TC is shown in Table 8.6, which is obtained with a more expensive daily operating cost of $126,413. However, the proposed TC approach provides a preventive UC schedule. Comparing the new schedule with that of Table 8.6, we learn that unit 2 is committed at hour 13 instead of hour 20, and unit 3 is not committed at hour 10. The flow violations at this hour are mitigated by switching off lines 2–4 and 4–5. The operating cost with TC is $126,271. The state of switchable lines before the outage of unit 2 is shown in Table 8.7. However, after the outage of unit 2, both switchable lines 2–4 and 4–5 will be switched off for the entire scheduling horizon.

Table 8.6 UC schedule of six-bus system in case 3 without TC.

Unit	Hours (0–24)																								
1	1	1	1	1	1	1	1	1	1	1	1	1	1	1	1	1	1	1	1	1	1	1	1	1	1
2	0	0	0	0	0	0	0	0	0	0	0	0	0	0	1	1	1	1	1	1	1	0	0	0	0
3	0	0	0	0	0	0	0	0	0	0	1	1	1	1	1	1	1	1	1	1	1	1	1	1	0

Table 8.7 Line schedule of six-bus system in case 3 using TC before outage of unit 2.

Line		Hours (0–24)																							
2–4	—	1	1	1	1	1	1	1	1	1	1	1	1	1	0	1	1	1	1	1	1	1	1	0	1
4–5	—	0	0	0	0	0	0	0	0	0	1	0	1	1	0	1	1	1	1	1	1	1	1	0	0

In these three cases, all lines are considered as switchable. However, only lines 2–4 and 4–5 are switched in these cases. In Figure 8.3, switching off any of lines 2–3, 3–6, 4–5, and 5–6 will remove the loop 2–3–4–5–6–2 with better results. Since line 4–5 has the lowest capacity, it is often subject to congestion, which makes the line a good candidate for switching. This conclusion is consistent with the proposed results. By switching off line 4–5, the line flow in the loop, i.e., those of lines 2–3, 3–6, and 5–6, can be raised without affecting the flow on other lines. Next, one of the lines in the other loop, i.e., lines 1–2, 1–4, and 2–4, could be a good choice for switching. Since unit 1 is the most economical and the largest unit, it will remain in service along with the attached lines. So, line 2–4 would be the best TC choice for this loop. With switching off both lines 2–4 and 4–5, enough lines are still connected to bus 4 to supply its load. By considering lines 2–4 and 4–5 as switchable, loops would be relaxed while meeting the load balance requirements.

Case 4: In this case, the standing phase angle difference limits of the switchable lines are considered. The standing phase angle difference limit for switchable lines is considered to be 4°. Imposing this limit, when the absolute value of the difference of phase angles on two sides of the switched line is less than its standing phase angle difference limit, the line cannot be switched back to the system.

In Case 1, switchable lines 2–4 and 4–5 are switched back to the system at hour 16. The standing phase angle differences for these two lines are 10.24° and 1.85°, respectively. Since the standing phase angle difference of line 2–4 is more than its limit, this line cannot be switched back at hour 16. Imposing this constraint to the problem, the switchable line 2–4 is switched back to the system one hour earlier, i.e., hour 15. The line can be switched back to the system at this hour since the standing phase angle difference is less than its limit. With this change in the state of switchable lines, the generation of unit 1 is decreased by 2.5 MW, and the generation of unit 2 is increased by 2.5 MW. This change in dispatch leads to an insignificant increase in the total operating cost.

8.3.2 IEEE 118-Bus System

A modified IEEE 118-bus system is used to study the SCUC with TC. The system has 118 buses, 54 units, and 186 branches. The data for this system is found in motor.ece.iit.edu/data/SCUC_118test.xls. Lines 30, 78, 90, 115, 151, 159, and 164 are considered switchable. Three cases are considered. The first case is the base case UC, and the second is SCUC when considering contingencies. In the third case, the results of the proposed model are compared with those of an integrated model.

Case 1: In this case, the SCUC approach without TC is used to find the generating unit schedules with a total operating cost of $1,081,320.36. Using TC, seven lines are considered switchable. The UC schedule found in the first iteration of the problem cannot satisfy the TC feasibility and Benders cuts are added to the UC problem. The feasible schedule is found after two iterations and is shown in Table 8.8. The TC schedule is shown in Table 8.9 with a total operating cost of $1,078,155.86, which signifies a 0.30% improvement in the total operating cost compared to the UC solution. The total execution time is 94 seconds.

Case 2: In this case, three simultaneous contingencies are considered. The contingencies include outages of unit 13, line 75–77, and line 85–89. Comparing the new schedule with that in Case 1, the states of twenty units have changed at different hours with a total operating cost of $1,081,898. To apply the proposed TC approach, the same lines as in Case 1 are considered switchable. The first UC result is feasible for the outage of line 75–77. So, the contingency of line 75–77 is controllable. Since the UC solution cannot lead to a feasible solution for the outages of unit 13 and line 85–89, these contingencies are uncontrollable. The obtained switchable line schedule is completely different from that in Case 1. The differences are highlighted in Table 8.9 compared to those in Case 1. The corresponding UC solution is shown in Table 8.8, where the highlighted values show the difference with Case 1. The total operating cost is $1,080,846, which shows a 0.09% improvement. This solution is obtained in 136 seconds.

Case 3: To show the effectiveness of the proposed model in handling a larger number of switchable lines, the TC cases with integrated and decomposed formulations are compared. The number of switchable lines is changed from 0 to 50 with a step of 10. Zero means that no switchable line exists in the system, while 50 represents that more than one-fourth of the transmission lines in the system are switchable. The default relative optimality gap of CPLEX, i.e., 0.01%, is used as the stopping criterion. The total operating cost and execution time for these two models for different numbers of switchable lines are obtained and shown in Figures 8.4 and 8.5, respectively.

In Figure 8.4, the decomposed model does not outperform the integrated model from an optimality point of view. The integrated model always has a similar or better total operating cost. However, the difference between the total operating costs of the two models is negligible. The two models result in the same solution when

Table 8.8 UC schedule of IEEE 118-bus system in case 1 using TC.

Unit	Hours (0–24)																							
1–2	0	0	0	0	0	0	0	0	0	0	0	0	0	0	0	0	0	0	0	0	0	0	0	0	0
3	0	0	0	0	0	0	0	0	0	0	0	0	0	0	0	0	0	0	0	0	0	0	0	0	0
4–5	1	1	1	1	1	1	1	1	1	1	1	1	1	1	1	1	1	1	1	1	1	1	1	1	1
6	0	0	0	0	0	0	0	0	0	0	0	0	0	0	0	0	0	0	0	0	0	0	0	0	0
7	0	0	0	0	0	0	0	0	0	0	1	1	1	1	1	1	1	1	1	1	1	1	1	1	0
8	0	0	0	0	0	0	0	0	0	0	0	0	0	0	0	0	1	1	0	0	0	0	0	0	0
9	0	0	0	0	0	0	0	0	0	0	0	0	0	0	0	1	1	1	0	0	0	0	0	0	0
10–11	1	1	1	1	1	1	1	1	1	1	1	1	1	1	1	1	1	1	1	1	1	1	1	1	1
12	0	0	0	0	0	0	0	0	0	0	0	0	0	0	0	0	0	1	0	0	0	0	0	0	0
13	0	0	0	0	0	0	0	0	0	0	0	0	0	0	0	0	0	1	0	0	0	0	0	0	0
14	0	0	0	0	0	0	0	0	0	0	1	1	1	1	1	1	1	1	1	1	1	1	1	1	0
15	0	0	0	0	0	0	0	0	0	0	0	0	0	0	0	0	0	1	0	0	0	0	0	0	0
16	0	0	0	0	0	0	0	0	0	0	1	1	1	1	1	1	1	1	1	1	1	1	1	0	0
17	0	0	0	0	0	0	0	0	0	0	0	0	0	0	0	0	1	1	0	0	0	0	0	0	0
18	0	0	0	0	0	0	0	0	0	0	0	0	0	0	0	0	1	1	0	0	0	0	0	0	0
19	0	0	0	0	0	0	0	0	0	0	1	1	1	1	1	1	1	1	1	1	1	1	1	1	0
20–21	1	1	1	1	1	1	1	1	1	1	1	1	1	1	1	1	1	1	1	1	1	1	1	1	1
22	0	0	0	0	0	0	0	0	0	0	1	1	1	1	1	1	1	1	1	1	1	1	1	0	0
23	0	0	0	0	0	0	0	0	0	0	1	1	1	1	1	1	1	1	1	1	1	1	1	0	0
24–25	1	1	1	1	1	1	1	1	1	1	1	1	1	1	1	1	1	1	1	1	1	1	1	1	1
26	0	0	0	0	0	0	0	0	0	0	1	1	1	1	1	1	1	1	1	1	1	1	1	0	0
27–29	1	1	1	1	1	1	1	1	1	1	1	1	1	1	1	1	1	1	1	1	1	1	1	1	1

(Continued)

Table 8.8 (Continued)

Unit	Hours (0–24)																								
30	0	0	0	0	0	0	0	0	0	0	0	1	1	1	1	1	1	1	1	1	1	1	1	0	0
31	0	0	0	0	0	0	0	0	0	0	0	0	0	0	0	1	1	1	0	0	0	0	0	0	0
32	0	0	0	0	0	0	0	0	0	0	0	0	0	0	0	0	1	1	0	0	0	0	0	0	0
33	0	0	0	0	0	0	0	0	0	0	0	0	0	0	0	0	0	0	0	0	0	0	0	0	0
34–35	0	0	0	0	0	0	0	0	0	0	1	1	1	1	1	1	1	1	1	1	1	1	1	1	1
36	1	1	1	1	1	1	1	1	1	1	1	1	1	1	1	1	1	1	1	1	1	1	1	1	1
37	0	0	0	0	0	0	0	0	0	1	1	1	1	1	1	1	1	1	1	1	1	1	1	1	1
38	0	0	0	0	0	0	0	0	0	0	0	0	0	0	0	0	0	0	0	0	0	0	0	0	0
39	1	1	1	1	1	1	1	1	1	1	1	1	1	1	1	1	1	1	1	1	1	1	1	1	1
40	0	0	0	0	0	0	0	0	0	0	0	0	0	1	1	1	1	1	1	1	1	1	1	0	0
41–42	0	0	0	0	0	0	0	0	0	0	0	0	0	0	0	0	0	0	0	0	0	0	0	0	0
43	0	0	0	0	0	0	0	0	1	1	1	1	1	1	1	1	1	1	1	1	1	1	1	1	1
44–45	1	1	1	1	1	1	1	1	1	1	1	1	1	1	1	1	1	1	1	1	1	1	1	1	1
46	0	0	0	0	0	0	0	0	0	0	0	0	0	0	0	0	0	0	0	0	0	0	0	0	0
47	0	0	0	0	0	0	0	0	0	0	0	0	0	1	1	1	1	1	1	1	1	1	1	0	0
48	0	0	0	0	0	0	0	0	0	0	0	0	0	1	1	1	1	1	1	1	1	1	1	0	0
49	0	0	0	0	0	0	0	0	0	0	0	0	0	0	0	0	0	0	0	0	0	0	0	0	0
50	0	0	0	0	0	0	0	0	0	0	0	0	0	0	0	0	0	0	0	0	0	0	0	0	0
51	0	0	0	0	0	0	0	0	0	0	0	0	1	1	1	1	1	1	1	1	1	1	1	0	0
52	0	0	0	0	0	0	0	0	0	0	0	0	1	1	1	1	1	1	1	1	1	1	1	0	0
53	0	0	0	0	0	0	0	0	0	0	1	1	1	1	1	1	1	1	1	1	1	1	1	1	0
54	0	0	0	0	0	0	0	0	0	0	0	0	0	0	0	1	1	1	0	0	0	0	0	0	0

Table 8.9 Line schedule in case 1 using TC.

Line		Hours (0–24)																							
30	—	1	1	0	0	0	1	1	0	0	0	0	1	0	1	0	0	0	0	0	1	1	0	1	1
78	—	0	1	0	0	0	1	1	0	0	0	0	1	1	0	0	0	0	0	0	1	1	0	1	1
90	—	1	1	0	0	0	1	1	0	0	0	1	1	1	1	1	1	1	1	1	0	0	0	1	1
115	—	0	1	1	1	1	1	1	0	0	1	0	1	0	1	1	0	1	0	0	1	1	1	1	1
151	—	1	1	1	1	1	1	1	0	0	0	1	1	0	0	1	1	1	1	1	0	0	1	1	1
159	—	1	0	0	0	0	0	1	1	0	0	1	1	1	1	1	1	1	1	1	1	1	1	1	1
164	—	0	0	0	0	0	0	0	0	0	0	0	0	0	0	0	0	0	0	0	0	0	0	0	0

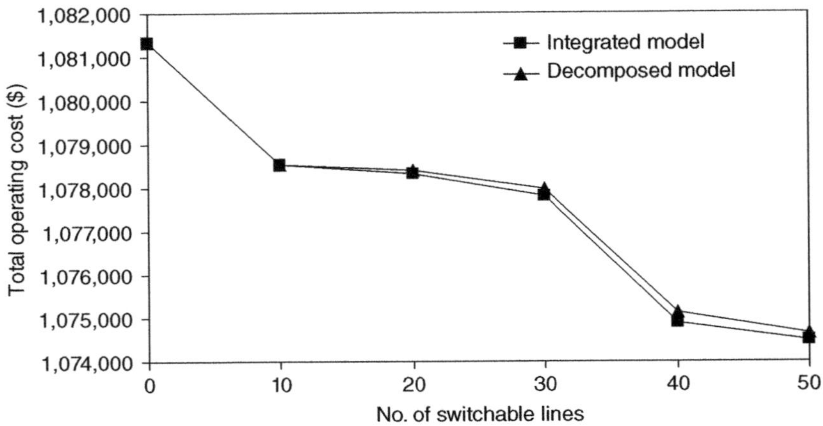

Figure 8.4 Total operating cost comparison of integrated and decomposed models.

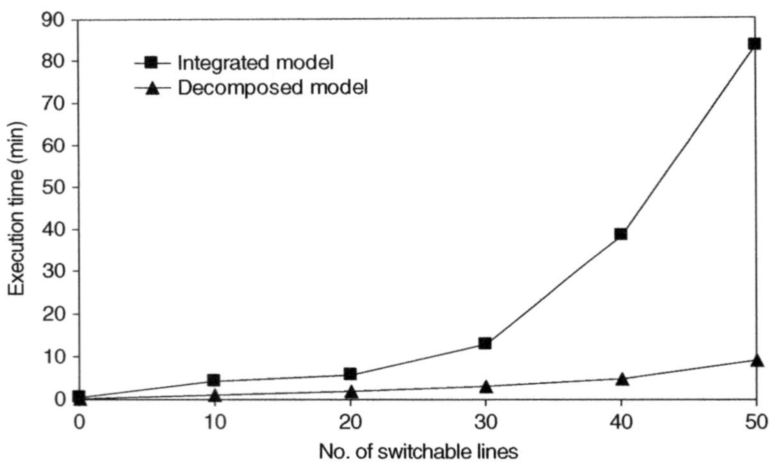

Figure 8.5 Execution time comparison of integrated and decomposed models.

there is no line switching. Since the proposed model is an extension of the decomposed SCUC model, it is expected that if there is no line switching, the decomposed model will find the same solution as the integrated model. The worst case occurs at 40 switchable lines, where the difference is 0.02%.

In Figure 8.5, for the small number of switchable lines, both models find the solution in a reasonable time. For instance, for ten switchable lines, the decomposed model finds the solution in one minute, and the integrated model finds the solution in four minutes. However, as we increase the number of switchable lines, the execution time of the integrated model increases significantly, while that of the decomposed model remains reasonable. In the case of 50 switchable lines, the solution of the decomposed model is obtained in nine minutes, while that of the integrated model is achieved in more than 80 minutes. This considerable increase in execution time is due to the large number of binary variables that the integrated problem should handle. However, in the decomposed model, these binary variables are divided between the UC problem and the TC problem. This division would help the large problems to find solutions in a reasonable execution time.

In the proposed model, both UC and the Optimal TC Scheduling blocks use the MIP format. MIP problems use a pre-defined gap to determine the optimality of the solution. If this gap is set to zero, the problem will find the optimal solution with a longer computation time. Otherwise, the problem will be terminated with a near-optimal solution in a reasonable time. Usually, a near-optimal solution is considered in real-time applications in power systems. The proposed TC approach could be further improved by applying the following:

- Numerical methods may be developed to identify the optimal set of switchable lines. Such methods may include additional strategies for system operations. For example, in the IEEE 118-bus system, line 164 would be a good prospect for switching since it is switched off periodically.
- Practical line-switching strategies could be considered, such as how often and how long a line could be on/off.
- Switching states of lines are employed in the Optimal TC Scheduling by means of cuts. These binary variables will be adjusted more accurately if we use additional constraints to make stronger connections between the switchable line states and the generation dispatch. The approach could help the TC problem find the optimal solution in fewer iterations [31].

8.4 Summary

In this chapter, TC was integrated with UC to solve the multi-interval optimal generation unit scheduling with security constraints. The Benders decomposition was employed to separate the problem into a UC problem and a TC problem. The UC problem minimizes the total operating cost and calculates the optimal hourly commitment and dispatch of generation units subject to power balance and reserve requirements. The TC problem adjusts the hourly unit schedules and optimizes the switching state of lines to mitigate transmission violations in the base case and controllable and uncontrollable contingencies. The features of the proposed approach are listed as follows: (i) the TC model can be used for congestion management. The TC application will lead to adjustments in line flows and congestion levels; (ii) the number of switchable lines was limited here. This assumption is consistent with practical switching applications. If we increase the number of switchable lines, TC may find better SCUC solutions, which could also converge at slower rates; (iii) the applications of Benders decomposition to SCUC and TC would make the proposed approach more applicable to large power systems; (iv) the proposed TC approach is an extension of the conventional SCUC. Hence, if we ignore the line switching in the proposed approach, the final SCUC solution will be the same as that of the conventional SCUC, (v) The proposed approach allows for the utilization of corrective modes before resorting to more expensive preventive solutions.

References

1 M. Shahidehpour, H. Yamin, and Z. Y. Li, *Market Operations in Electric Power Systems*. New York, NY, USA: Wiley, 2002.

2 A. J. Conejo, E. Castillo, R. Mínguez, and R. García-Bertrand, *Decomposition Techniques in Mathematical Programming*. Berlin, Germany: Springer, 2006.

3 R. Bacher and H. Glavitsch, "Network topology optimization with security constraints," *IEEE Trans. Power Syst.*, vol. 1, pp. 103–111, Nov. 1986.

4 G. Schnyder and H. Glavitsch, "Integrated security control using an optimal power flow and switching concepts," *IEEE Trans. Power Syst.*, vol. 3, pp. 782–790, May 1988.

5 G. Schnyder and H. Glavitsch, "Security enhancement using an optimal switching power flow," *IEEE Trans. Power Syst.*, vol. 5, pp. 674–681, May 1990.

6 J. G. Rolim and L. J. B. Machado, "A study of the use of corrective switching in transmission systems," *IEEE Trans. Power Syst.*, vol. 14, pp. 336–341, Feb. 1999.

7 W. Shao and V. Vittal, "Corrective switching algorithm for relieving overloads and voltage violations," *IEEE Trans. Power Syst.*, vol. 20, no. 4, pp. 1877–1885, Nov. 2005.

8 G. Granelli, M. Montagna, F. Zanellini, P. Bresesti, R. Vailati, and M. Innorta, "Optimal network reconfiguration for congestion management by deterministic and genetic algorithms," *Electr. Power Syst. Res.*, vol. 76, pp. 549–556, Apr. 2006.

9 R. P. O'Neill, R. Baldick, U. Helman, M. H. Rothkopf, and W. Stewart, "Dispatchable transmission in RTO markets," *IEEE Trans. Power Syst.*, vol. 20, no. 1, pp. 171–179, Feb. 2005.

10 E. B. Fisher, R. P. O'Neill, and M. C. Ferris, "Optimal transmission switching," *IEEE Trans. Power Syst.*, vol. 23, no. 3, pp. 1364–1355, Aug. 2008.

11 K. W. Hedman, R. P. O'Neill, E. B. Fisher, and S. S. Oren, "Optimal transmission switching with contingency analysis," *IEEE Trans. Power Syst.*, vol. 24, no. 3, pp. 1577–1586, Aug. 2009.

12 K. W. Hedman, R. P. O'Neill, E. B. Fisher, and S. S. Oren, "Optimal transmission switching—sensitivity analysis and extensions," *IEEE Trans. Power Syst.*, vol. 23, no. 3, pp. 1469–1479, Aug. 2008.

13 P. Dehghanian and M. Kezunovic, "Probabilistic decision making for the bulk power system optimal topology control," *IEEE Trans. Smart Grid*, vol. 7, no. 4, pp. 2071–2081, Jul. 2016, doi: https://doi.org/10.1109/TSG.2016.2544794.

14 A. Nikoobakht, M. Mardaneh, J. Aghaei, V. G. Mestre, and J. Contreras. "Flexible power system operation accommodating uncertain wind power generation using transmission topology control: an improved linearised AC SCUC model," *IET Gener. Transm. Distrib.*, vol. 11, no. 1, pp. 142–153, 2017.

15 R. Roychowdhury, J. B. Ocampo, B. Guddanti, and M. S. Illindala, "Substation topology and line switching control using deep reinforcement learning," in *2022 IEEE/IAS 58th Ind. Commer. Power Syst. Techn. Conf. (I&CPS)*, Las Vegas, NV, USA, 2022, pp. 1–6, doi: https://doi.org/10.1109/ICPS54075.2022.9773937.

16 J. Toctaquiza, D. Carrión, and M. Jaramillo, "An electrical power system reconfiguration model based on optimal transmission switching under scenarios of intentional attacks," *Energies*, vol. 16, no. 6, p. 2879, 2023.

17 Y. Fu, M. Shahidehpour, and Z. Li, "Security-constrained unit commitment with AC constraints," *IEEE Trans. Power Syst.*, vol. 20, no. 3, pp. 1538–1550, Aug. 2005.

18 M. Shahidehpour and Y. Fu, "Benders decomposition," *IEEE Power Energy Mag.*, vol. 3, no. 2, pp. 20–21, Mar. 2005.

19 M. Shahidehpour and V. Ramesh, "Nonlinear programming algorithms and decomposition strategies for OPF," in *IEEE/PES Tutorial Optimal Power Flow*. Piscataway, NJ: IEEE Press, 1996.

20 H. Ma and M. Shahidehpour, "Transmission constrained unit commitment based on benders decomposition," *Elect. Power Energy Syst.*, vol. 20, no. 4, pp. 287–294, Apr. 1998.

21 A. M. Geoffrion, "Generalized benders decomposition," *J. Optim. Theory Appl.*, vol. 10, no. 4, pp. 237–261, 1972.

22 M. Adibi and R. Kafka, "Power system restoration issues," *IEEE Comput. Appl. Power*, vol. 4, no.2, pp. 19–24, Apr. 1991.

23 A. Ketabi and A. Ranjbar, "New approach to standing phase angle reduction for power system restoration," *IEEE Elect. Power Eng*. Budapest, Hungary: PowerTech, 1999.

24 N. Martins, E. J. Oliviera, W. C. Moreira, J. L. R. Pereira, and R. M. Fontoura, "Redispatch to reduce rotor shaft impacts upon transmission loop closure," *IEEE Trans. Power Syst.*, vol. 23, no. 2, pp. 592–600, May 2008.

25 T. Nagata, H. Sasaki, and R. Yokoyama, "Power system restoration by joint usage of expert system and mathematical programming approach," *IEEE Trans. Power Syst.*, vol. 10, no. 3, pp. 1473–1479, Aug. 1995.

26 S. Wunderlich, M. M. Adibi, R. Fischl, and C. O. D. Nwankpa, "An approach to standing phase angle reduction," *IEEE Trans. Power Syst.*, vol. 9, no. 1, pp. 470–476, Feb. 1994.

27 D. Hazarika and A. K. Sinhá, "Standing phase angle reduction for power system restoration," *Proc. Inst. Elect. Eng. Gen. Transm. Distrib.*, vol. 145, no. 1, pp. 82–88, Jan. 1998.

28 D. Hazarika and A. K. Sinhá, "An algorithm for standing phase angle reduction for power system restoration," *IEEE Trans. Power Syst.*, vol. 14, no. 4, pp. 1213–1218, Nov. 1999.

29 N. Martins, E. J. de Oliveira, J. L. R. Pereira, and L. C. A. Ferreira, "Reducing standing phase angles via interior point optimum power flow for improved system restoration," in *Proc. Power Systems Conf. Expo. 2004*, New York, NY, USA, Oct. 10–13, 2004.

30 ILOG CPLEX, "ILOG CPLEX homepage." 2009. Accessed: Dec 2009. [Online]. Available: http://www.ilog.com

31 J. Guo and Z. Li, "Network reduction with unknown transmission line status," in *39th North American Power Symp.*, 2007.

9

The Economics of Decentralization Through Microgrids

9.1 Principles of Microgrids

Microgrids, as small-scale power systems with the ability to self-supply and island, have emerged as core components of the grid of the future [1–3]. Microgrids are viable solutions to host Distributed Energy Resources (DERs) and facilitate their growing penetration in the distribution grids [4–7] while at the same time improving the power systems' reliability and resiliency, enhance the individuals' economic benefits, and support the penetration of renewable energy resources in distribution grids [8–11]. Microgrids can disconnect from the utility grid during upstream events to improve load-point reliability, besides working in grid-connected mode to improve economic objectives [12–16]. Specifically, microgrids that are geographically close to each other can be electrically connected to construct a dynamic structure that can further expand the projected benefits and potentially reduce the impact of disasters [17–23].

Controlling the operation of the integrated microgrids in a centralized approach is not most suitable for microgrids due to privacy concerns. On the other hand, current distributed solutions may suffer from convergence issues and require peer-to-peer communication infrastructure [24–27]. Consequently, developing a more intelligent technique that balances these two approaches for managing the integrated microgrid operation seems significant in pursuing efforts to achieve the desired resilience objectives.

An adaptive distribution grid can support microgrid integration and mitigate the challenges above [27–29]. In an adaptive distribution grid, microgrids are integrated within a static framework, and their behavior is prescribed by a supervisory controller. Therefore, the system autonomy is preserved, and the overall system of systems is managed by a supervisory controller. The adaptive distribution grid is expected to be an essential application in future smart grids [27, 30] due to its privilege in (i) enhancing the information and power exchange among integrated

The Economics of Climate Resilience in Power Infrastructure, First Edition. Ali Arabnya and Amin Khodaei.

microgrids, (ii) optimizing the aggregated system performance by offering the capability of a dynamic reconfiguration, (iii) fostering the diversity of energy resources, system autonomy, and the connectivity among integrated systems, and (iv) balancing individual microgrid's and integrated microgrid's objectives.

Clustering integrated microgrids in the adaptive distribution grid creates distinct levels of integrated microgrids, called super-microgrids, for enhancing individual- and aggregated-system objectives. It is worth mentioning that the generated super-microgrids can be reconfigured or reorganized to form different super-microgrids based on certain optimization criteria [27, 31]. This architecture is limitedly discussed and investigated in the literature, primarily under the context of holonic distribution grids. The study in [32] proposes a generic architecture system based on the holarchy concept for smart grids, where the proposed architecture comprises multiple autonomous prosumers that are recursively interconnected at various combination layers with a bottom-up organization. In addition, the control functions of the system are modeled based on the service-oriented architecture framework to control the system. In [33], the smart grid control challenges are considered by proposing a holonic control architecture that contributes to achieving local and global objectives, i.e., at the macro and micro levels. The proposed holonic architecture system is examined by exemplifying a proof-of-concept implementation by integrating basic control solutions. In [34], a goal-based holonic multi-agent system is proposed to support the operation of the power distribution systems as cyber-physical systems. In addition, it discusses the contribution of the holonic multi-agent system for two vital applications in the power distribution systems involving the system state estimation provided by smart meters installed at homes and the control of the reactive power at rooftop PV installations.

A holonic multi-agent system architecture is presented in [35] for adaptive control of the distribution grids. The proposed architecture can optimize the system performance and maintain the operation within limits. In addition, the proposed study takes the system's islanded operation into account in emergencies. The study in [36] presents a fractal model based on a holonic structure for power grids. The presented model manages the system in a bottom-up way, allowing the aggregation of multiple layers and sufficiently maintaining the involved systems' autonomy by enabling the local management of the installed resources. In [37], an intelligent control strategy based on a holonic system structure for renewable energy resources is presented, where the proposed holonic structure includes resource holon, energy holon, and service holon. In addition, a multi-criteria decision aid method is performed to control the system components. The study in [30] presents an optimal scheduling model of integrated microgrids in the holonic distribution grids, where the proposed model can identify the optimal network topology that minimizes holon-specific and system-wide operation costs and maximizes the overall system reliability. The proposed model is formulated using mixed-integer programming and numerically simulated via a holonic distribution grid test system with multiple microgrids. In [31], the impact of the microgrids' spinning reserve in enabling the power exchange among the integrated microgrids in the holonic distribution grids is discussed and investigated. In addition, an illustrative study is performed to examine the influence of the individual holon's spinning reserve on forming optimal super-holons combinations that can improve participated holons' objectives. Ref. [38] analyzed the decentralized economic scheduling of multi-microgrids, considering the uncertainty in wind power and solar photovoltaic outputs. A decoupling method for public connection points to partition multi-microgrid regions was introduced. Additionally, a decentralized economic dispatch model to minimize economic costs while managing the uncertainties of renewable energy outputs was proposed. The authors employed an Alternating Direction Method of Multipliers (ADMM) algorithm to solve the scheduling problem. Simulation tests and comparative analyses validate the feasibility and superiority of this algorithm. Authors in [39] investigated the strength of the power grid to withstand extreme weather and natural disasters by suggesting the implementation of prosumer-centric networked microgrids. These microgrids combine various DERs such as solar PV panels, wind turbines, battery energy storage systems (BESSs), and electric vehicles (EVs) to form self-sufficient units capable of functioning independently during grid disruptions. A sequential Monte Carlo analysis was used to optimize the allocation of DERs and EVs with the aim of improving system resilience. The findings highlight the advantages of prosumer-centric microgrids for energy resilience. Ref. [40] investigated the application of quantum computing to improve the distribution grid resilience by creating microgrids to restore critical loads after natural disasters or faults. A quantum-inspired optimization model that uses quantum annealing to tackle complex combinatorial problems was proposed. The study utilizes the Constrained Quadratic Model (CQM) solver from D-Wave, which combines classical and quantum optimization techniques to deliver high-performance solutions. The effectiveness of the quantum approach was confirmed through a comparative analysis with the classical Gurobi solver. The results demonstrate the high potential of quantum computing in handling large-scale optimization problems in power systems restoration and improvement of system resilience.

This chapter proposes a spinning reserve-based optimal scheduling model of integrated microgrids in an adaptive distribution grid to address common resilience issues in the face of disasters. This model solves the common convergence issues with the existing models in the literature by moving away from a power exchange-focused modeling and convergence and adopting a reserve-based approach. The developed model aims to identify the optimal super-microgrid combinations based on minimum net spinning reserve. Identifying optimal super-microgrids could improve participating players' economic benefits and enhance system resilience.

The rest of this chapter is organized as follows. The proposed model is outlined in Section 9.2, and the problem is mathematically formulated in Section 9.3. Numerical simulations are provided in Section 9.4, and Section 9.5 concludes the chapter.

9.2 Economic Variables for Decentralized Power Systems

Consider an adaptive distribution grid that consists of several microgrids. As depicted in Figure 9.1, in the normal operation mode of the grid, microgrids are connected to the utility grid and operated in the grid-connected mode. In this mode, each microgrid aims to determine the least-cost commitment and dispatch of available dispatchable units, charging/discharging schedules of energy storage systems, schedule of adjustable loads, and the exchanged power with the utility grid. Owing to economic discretion, each microgrid in the grid-connected mode prefers to exchange power with the utility grid rather than with adjacent microgrids. When it comes to the islanded operation mode, however, the microgrids in the adaptive distribution grid are disconnected from the utility grid. Accordingly, each microgrid relies on its local resources and the spinning reserve provided by the adjacent microgrid to meet its load demands. In case of lacking adequate local capacity in a microgrid, i.e., power deficiency, microgrids can make connections based on the spinning reserve to form integrated microgrids and accordingly increase the microgrid capacity, as well as overall system reliability and resilience. Nevertheless, as the spinning reserve provision is costly, the configuration of connecting microgrids is optimally determined via a spinning reserve-based integrated microgrids scheduling model as proposed in this chapter.

The proposed spinning reserved-based optimal scheduling model for integrated microgrids is cast to minimize not only the operation cost associated with all microgrids in the grid-connected operation but also the costs of power deficiency and spinning reserve in the islanded operation mode. Since the Value of Lost Load (VOLL) is significantly higher than the value of the net spinning reserve, it is more economical to reap the benefits of the system-aggregated spinning reserve and make the system reliable and resilient by reducing the amount of power deficiency. Thus, determining an optimal configuration of the system in the islanded operation, i.e., optimal super-microgrids combination, plays a key role in minimizing the system-aggregated operation cost and improving the overall system reliability and resilience.

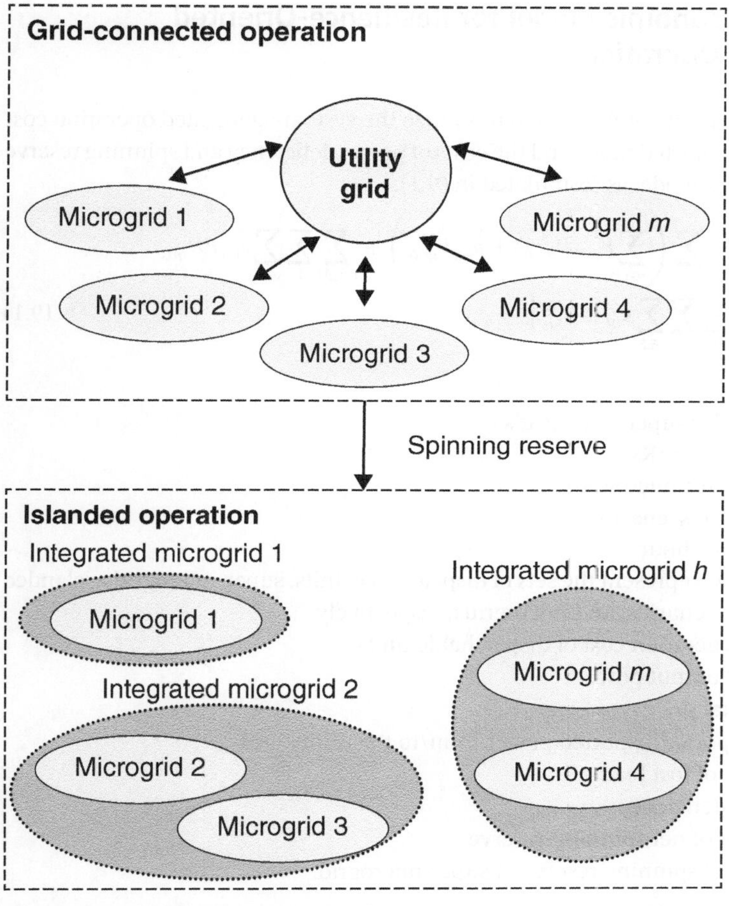

Figure 9.1 Proposed spinning reserve-based model for integrated microgrids in an adaptive distribution grid.

The spinning reserve for each microgrid is characterized based on the dispatchable DGs and energy storage systems available in that microgrid. Given that each microgrid could experience power deficiency during the islanded operation, the net spinning reserve is determined for the microgrid. Each super-microgrid is formed so that the net spinning reserve of that specific super-microgrid is minimized. In other words, among all possible microgrid combinations to form super-microgrids, those combinations that provide the least net spinning reserve during the islanded operation are the desired ones that are determined through the proposed model.

9.3 An Economic Model for Resilience-Oriented Microgrid Operations

The objective of the problem is to minimize the system-aggregated operation cost in the grid-connected mode and the costs of power deficiency and spinning reserve in the islanded mode, as formulated in (9.1):

$$min \sum_{m\in M}\sum_{t}\left(\sum_{i\in G} F_{mi}(P_{mit0}) + \rho_{mt}P^{M}_{mt0}\right) + \sum_{m\in M}\sum_{t}\sum_{s\in K} v_{mt}PD_{mts} + \sum_{h\in H}\sum_{t}\sum_{s\in K}\pi_{ht}\left|SR^{Net}_{hts}\right| \tag{9.1}$$

where

h is an index for super-microgrids
i is an index for DERs
m is an index for microgrids
s is an index for scenarios
t is an index for hour
G, H, K, and M represent the sets of dispatchable units, super-microgrids, islanded operation scenarios, and microgrids, respectively
$F(\bullet)$ is the generation cost of dispatchable units
P is the DER output power
ρ is the market price
P^M is the imported/exported power from/to the utility grid
v is the value of lost load
PD is power deficiency
π is the value of net spinning-reserve
SR^{net} is the net spinning reserve of super-microgrid.

This objective function consists of four distinct terms, which include the generation cost of all dispatchable DGs related to each microgrid, the cost of exchanged power between each microgrid and the utility grid, the cost of power deficiency (which will be translated to load curtailment if additional generation is not available) as well as the cost of spinning reserve in the islanded operation. The generation cost is often formulated via a quadratic function; however, it is linearized through a piecewise linearization approach. The second term is the cost of exchanged power with the utility grid, calculated based on the market price at the Point of Interconnection (POI) for each microgrid. This term could be either positive (denoting a cost) or negative (denoting a revenue), according to the power flow direction associated with each microgrid. The third term represents the cost of power deficiency for each microgrid during the islanded operation. This term is calculated as the VOLL multiplied by the deficient power in each microgrid. The last term indicates the cost of the net spinning reserve of each super-microgrid, calculated based on the multiplication of the value of the net spinning reserve and the absolute value for the net spinning reserve of that super-microgrid. The absolute value causes nonlinearity, which is further linearized. The absolute value is assigned to an auxiliary variable ϕ_{hts}, i.e., $\phi_{hts} = \left|SR_{hts}^{Net}\right|$. Next, the objective is to minimize ϕ_{hts}, subject to an additional constraint, i.e., $-\phi_{hts} \leq SR_{hts}^{Net} \leq \phi_{hts}$.

In the proposed problem, two operation modes, i.e., grid-connected and islanded, are investigated. These two modes are separated from each other in the objective function and further in the constraints using scenario index s, where $s = 0$ is used for the grid-connected mode and $s \geq 1$ for the islanded mode. This objective is subject to prevailing operational and spinning reserve constraints, listed as follows:

Microgrid DG constraints:

$$\sum_{i\in R} P_{mit} + P_{mt}^{M} = D_{mt}, \quad \forall m, \forall t \tag{9.2}$$

$$-P_{m}^{M,max} U_{ts} \leq P_{mts}^{M} \leq P_{m}^{M,max} U_{ts}, \quad \forall m, \forall t, \forall s \tag{9.3}$$

$$P_{mi}^{min} I_{mit} \leq P_{mits} \leq P_{mi}^{max} I_{mit}, \quad \forall m, \forall i \in G, \forall t, \forall s \tag{9.4}$$

$$P_{mits} - P_{mi(t-1)s} \leq UR_{mi}, \quad \forall m, \forall i \in G, \forall t, \forall s \tag{9.5}$$

$$P_{mi(t-1)s} - P_{mits} \leq DR_{mi}, \quad \forall m, \forall i \in G, \forall t, \forall s \tag{9.6}$$

$$T_{mi}^{on} \leq UT_{mi}(I_{mit0} - I_{mi(t-1)}), \quad \forall m, \forall i \in G, \forall t \tag{9.7}$$

$$T_{mi}^{off} \geq DT_{mi}(I_{mi(t-1)} - I_{mit}), \quad \forall m, \forall i \in G, \forall t \tag{9.8}$$

where

D is the load demand
U represents the islanding state (0 when islanded, 1 otherwise)
I is the commitment state of dispatchable units (1 when committed, 0 otherwise)
UR and DR, respectively represent ramp-up and ramp-down rates
UT and DT, respectively represent minimum up time and minimum down time
T^{on} and T^{off}, respectively represent the number of successive *ON* hours and the number of successive *OFF* hours.

The load balance equation (9.2) ensures that the sum of power generated by all DERs (i.e., dispatchable and non-dispatchable units and energy storage systems) and the exchanged power with the utility grid would match the hourly load for each microgrid. This constraint is only considered for the grid-connected mode, and as each microgrid has adequate generation during this mode (either locally generated by DERs or imported from the utility grid), the power deficiency variable is not added to the equation. The generation of non-dispatchable units is forecasted on a day-ahead basis for each microgrid and then accordingly utilized in (9.2). The power associated with energy storage systems is either positive (discharging), negative (charging), or zero (idle). The power exchange between each microgrid and the utility grid can be positive (purchasing power), negative (selling back), or zero. This power exchange is limited by the thermal limit of the line connecting each microgrid to the utility grid (9.3). The binary islanding indicator U_{ts} ensures that the exchanged power with the utility grid is forced to be zero during the islanded operation. The dispatchable unit generation in each microgrid is restricted by the minimum and maximum generation capacity limits (9.4). The binary variable I_{mit} indicates the unit commitment state of that unit, which is one once committed and zero otherwise. Constraints (9.5) and (9.6) are, respectively, defined for ramp-up and ramp-down limits associated with dispatchable units in each microgrid. The minimum up and down time limits are formulated in (9.7) and (9.8), respectively.

Microgrid energy storage constraints:

$$P_{mits} \leq P_{mit}^{dch,max} u_{mit} - P_{mit}^{ch,min} v_{mit}, \quad \forall m, \forall i \in S, \forall t, \forall s \tag{9.9}$$

$$P_{mits} \geq P_{mit}^{dch,min} u_{mit} - P_{mit}^{ch,max} v_{mit}, \quad \forall m, \forall i \in S, \forall t, \forall s \tag{9.10}$$

$$u_{mit} + v_{mit} \leq 1, \quad \forall m, \forall i \in S, \forall t \tag{9.11}$$

$$C_{mits} = C_{mi(t-1)s} - \frac{P_{mits} u_{mits} \tau}{\eta_i^{dch}} - \eta_i^{ch} P_{mits} v_{mits} \tau, \quad \forall m, \forall i \in S, \forall t, \forall s \tag{9.12}$$

$$C_{mi}^{min} \leq C_{mits} \leq C_{mi}^{max}, \quad \forall m, \forall i \in S, \forall t, \forall s \tag{9.13}$$

$$T_{mi}^{ch} \geq MC_{mi}(u_{mit} - u_{mi(t-1)}), \quad \forall m, \forall i \in S, \forall t \tag{9.14}$$

$$T_{mi}^{dch} \geq MD_{mi}(v_{mit} - v_{mi(t-1)}), \quad \forall m, \forall i \in S, \forall t \tag{9.15}$$

where

ch is the superscript for energy storage charging mode
dch is the superscript for energy storage discharging mode
S is the set of energy storage systems
u is the energy storage discharging state (1 when discharging, 0 otherwise)
v is the energy storage charging state (1 when charging, 0 otherwise)
C is the energy storage stored energy
τ is the time period
η is the energy storage efficiency

MC is the minimum charging time
MD is the minimum discharging time
T^{ch} is the number of successive charging hours
T^{dch} is the number of successive discharging hours.

The energy storage system constraints mainly include limitations associated with its power and energy, charging/discharging mode, and minimum charging/discharging time. Constraints (9.9) and (9.10) characterize the energy storage system power for each microgrid based on the maximum and minimum charging and discharging power limits and operation mode. The energy storage system can be charging, discharging, or idle at each hour (9.11). The stored energy at each hour is determined according to the stored energy at the previous hour, the amount of charged/discharged power, and charging/discharging efficiency (9.12), which is also further restricted with minimum and maximum capacity limits (9.13). Constraints (9.14) and (9.15) respectively outline minimum charging and discharging time limits, denoting the minimum number of successive hours that the energy storage system must maintain charging or discharging once the operational mode is toggled.

Microgrid adjustable load constraints:

$$D^{min}_{mdt} z_{mdt} \leq D_{mdts} \leq D^{max}_{mdt} z_{mdt}, \quad \forall m, \forall d \in D_A, \forall t, \forall s \tag{9.16}$$

$$T^{on}_{md} \geq MU_{md}(z_{mdt} - z_{md(t-1)}), \quad \forall m, \forall d \in D_A, \forall t \tag{9.17}$$

$$\sum_{[\alpha_d, \beta_d]} D_{mdts} = E_{md}, \quad \forall m, \forall d \in D_A, \forall s \tag{9.18}$$

where

d is the index for loads
D_A is the set of adjustable loads
z is the adjustable load state (1 when operating, 0 otherwise)
E is the adjustable load total required energy
MU is the minimum operating time
T^{on} is the number of successive ON hours
α, β are the specified start and end times of adjustable loads.

Adjustable loads associated with each microgrid are bounded between minimum and maximum rated power limits (9.16). The minimum operating time (9.17) and the required energy to wrap up the operating cycle (9.18) are also regarded for adjustable loads.

Microgrid spinning reserve constraints:

$$PD_{mts} = D_{mts} - \sum_{i \in R} P_{mits}, \quad \forall m, \forall t, \forall s \tag{9.19}$$

$$sr^{DG}_{mits} \leq P^{max}_{mi} - P_{mits}, \quad \forall m, \forall i \in G, \forall t, \forall s \tag{9.20}$$

$$0 \leq sr^{DG}_{mits} \leq 10MSR_{mi}I_{mit}, \quad \forall m, \forall i \in G, \forall t, \forall s \tag{9.21}$$

$$0 \leq sr^{ES}_{mits} \leq P^{dch,max}_{mit} u_{mit}, \quad \forall m, \forall i \in S, \forall t, \forall s \tag{9.22}$$

$$sr^{ES}_{mits} \leq P^{dch,max}_{mit} - P_{mits}, \quad \forall m, \forall i \in S, \forall t, \forall s \tag{9.23}$$

$$sr^{net}_{mts} = \sum_{i \in G} sr^{DG}_{mits} + \sum_{i \in S} sr^{ES}_{mits} - PD_{mts}, \quad \forall m, \forall t, \forall s \tag{9.24}$$

where

$10MSR$ is the 10-minute maximum sustained rate
sr^{DG} is the spinning reserve of dispatchable DGs
sr^{ES} is the spinning reserve of the energy storage system
sr^{net} is the net spinning reserve of the super-microgrid.

Spinning reserve is one of the key resources utilized to ensure integrated microgrids' reliability in responding to unforeseen events such as islanding. In the case of islanding, each microgrid is disconnected from the utility grid, in which adequate generation capacity may not be available to fully supply local loads. The amount of deficient power for each microgrid during the islanded operation is calculated based on the difference between the microgrid load and the power generated by all DERs (9.19). The spinning reserve of a dispatchable DG cannot exceed the difference between its maximum generation capacity and current generation (9.20). This spinning reserve is also limited by the 10-minute maximum sustained rate (9.21). Similar to dispatchable DGs, energy storage systems can provide a spinning reserve to protect integrated microgrids in case of islanding. An energy storage system can contribute to spinning reserve only during discharging mode, which is restricted by its maximum discharging power limit (9.22). The spinning reserve of an energy storage system cannot be greater than the difference between its maximum discharging power and existing discharged power (9.23). The net spinning reserve of a microgrid is calculated based on the aggregated spinning reserve associated with dispatchable DGs and energy storage systems, and the power deficiency (9.24).

Super-microgrid spinning reserve constraints:

$$SR^{Net}_{hts} = \sum_{m \in M} \left(w_{mhts} sr^{net}_{mts} \right), \quad \forall h \in H, \forall t, \forall s \tag{9.25}$$

$$\sum_{h \in H} w_{mhts} = 1, \quad \forall m, \forall t, \forall s \tag{9.26}$$

$$\sum_{m \in M} \sum_{h \in H} w_{mhts} = N_m, \quad \forall t, \forall s \tag{9.27}$$

In the proposed adaptive distribution grid, it is assumed that microgrids are connected based on the available net spinning reserves (i.e., sr^{net}) to form super-microgrids. Binary variable w_{mhts} is defined to decide whether microgrid m belongs to super-microgrid $h$or not. If w_{mhts} is one, it means that microgrid m belongs to super-microgrid h. The number of super-microgrids is set to be equal to the number of microgrids, i.e., $N_m = N_h$. This means if an adaptive distribution grid comprises N_m microgrids, N_h (which equals to N_m) super-microgrids are formed out of those microgrids; however, there could be certain super-microgrids formed by no microgrids (empty member). The net spinning reserve of a super-microgrid (i.e., SR^{net}) is determined according to the net spinning reserve of those microgrids belonging to that super-microgrid (9.25). Constraint (9.26) ensures that each microgrid belongs to only one super-microgrid. To ensure that all microgrids are assigned to create super-microgrids, (9.27) is defined. Multiplication of variables w_{mhts} and sr^{net} makes (9.25) nonlinear, which is further linearized in (9.28)–(9.30). κ and B are an auxiliary variable and a large positive number, respectively.

$$-B(1 - w_{mhts}) \leq \kappa_{mhts} - sr^{net}_{mts} \leq B(1 - w_{mhts}), \quad \forall m, \forall h, \forall t, \forall s \tag{9.28}$$

$$-Bw_{mhts} \leq \kappa_{mhts} \leq Bw_{mhts}, \quad \forall m, \forall h, \forall t, \forall s \tag{9.29}$$

$$SR^{Net}_{hts} = \sum_{m \in M} \kappa_{mhts}, \quad \forall h \in H, \forall t, \forall s \tag{9.30}$$

9.4 Case Study

An adaptive distribution grid with five microgrids is employed to investigate the performance and effectiveness of the proposed spinning reserve-based optimal scheduling of integrated microgrids. Microgrids' characteristics are tabulated in Tables 9.1–9.5, which show the aggregated generation of non-dispatchable units, hourly fixed load, adjustable load characteristics, and the characteristics associated with dispatchable DGs and DES units. Table 9.6 demonstrates the hourly forecasted market price. The power exchange between the utility grid and each microgrid is limited by the connecting line limit, which is set to 15 MW for each of the five microgrids. The proposed problem is solved for a 24-hour scheduling horizon considering a one-hour time period, i.e., $\tau = 1$. A total of 25 scenarios are considered in this study (scenario 0 for the grid-connected and scenarios 1–24 for the islanded operation). Note that each islanding scenario denotes the islanded operation in a certain one-hour time interval during the 24-hour scheduling horizon. The problem is modeled and solved in a computer with Intel Core i5 2.3 GHz processor and 4 GB RAM using CPLEX 11.0 [41]. The following cases are studied:

Table 9.1 Aggregated generation of non-dispatchable units (MW).

Time (h)	1	2	3	4	5	6	7	8
MG 1	0	0	0	0	0.63	0.80	0.62	0.71
MG 2	0	0	0	0	0	0	0	0.02
MG 3	0.09	0.08	0.11	0.08	0.40	0.42	0.51	0.59
MG 4	0.04	0.05	0.08	0.09	0.01	0	0	0.01
MG 5	0	0	0	0	0	0	0.01	0.02
Time (h)	**9**	**10**	**11**	**12**	**13**	**14**	**15**	**16**
MG 1	0.68	0.35	0.62	1.11	1.21	1.57	1.23	1.28
MG 2	0.08	0.26	0.48	0.74	0.92	0.99	0.97	0.91
MG 3	0.69	0.87	0.91	0.93	0.90	0.94	0.92	0.95
MG 4	0.06	0.13	0.24	0.36	0.43	0.46	0.49	0.41
MG 5	0.09	0.29	0.51	0.57	0.65	0.61	0.53	0.44
Time (h)	**17**	**18**	**19**	**20**	**21**	**22**	**23**	**24**
MG 1	1.05	0.82	0.71	0.92	0.57	0.60	0	0
MG 2	0.83	0.72	0.45	0.12	0	0	0	0
MG 3	0.79	0.69	0.31	0.14	0.1	0.02	0.12	0.09
MG 4	0.32	0.19	0	0	0	0	0	0
MG 5	0.30	0.11	0.09	0.01	0	0	0	0

Case 1: Individual microgrid optimal scheduling.
Case 2: Spinning reserve-based optimal scheduling of integrated microgrids.

Case 1: Each of the five microgrids in this case is individually scheduled for both the grid-connected and islanded operation modes. In the grid-connected scenario, each microgrid determines the optimal schedule of its local resources as well as the exchanged power with the utility grid. In the islanded scenarios in this case, each microgrid only relies on its local resources, which means no spinning reserve is used by adjacent microgrids to support the system. The results show a total load curtailment of 10.08 MWh for all five microgrids.

Case 2: In this case, the microgrids are responsible for minimizing their operation costs and providing spinning reserves to the adjacent microgrids with power deficiency. The five microgrids aim to supply their own local loads while offering a spinning reserve to the adjacent microgrids to minimize the system-aggregated

Table 9.2 Hourly fixed load (MW).

Time (h)	1	2	3	4	5	6	7	8
MG 1	8.73	8.54	8.47	9.03	8.79	8.81	10.12	10.93
MG 2	6.20	6.19	6.07	5.91	4.43	4.79	5.09	4.75
MG 3	17.81	18.29	18.58	18.95	19.21	19.89	19.99	19.82
MG 4	5.29	6.16	6.34	5.79	5.99	6.11	5.84	4.99
MG 5	4.17	4.51	4.82	5.29	5.19	5.74	5.86	5.85
Time (h)	**9**	**10**	**11**	**12**	**13**	**14**	**15**	**16**
MG 1	11.19	11.78	12.08	12.13	13.92	15.27	15.36	15.69
MG 2	4.93	5.69	4.91	5.79	6.92	7.81	8.09	8.08
MG 3	19.56	19.11	18.16	18.27	17.63	16.31	16.12	15.09
MG 4	4.92	4.89	4.81	4.98	4.71	4.65	4.82	4.73
MG 5	6.47	6.72	6.74	6.81	7.49	7.24	8.11	8.64
Time (h)	**17**	**18**	**19**	**20**	**21**	**22**	**23**	**24**
MG 1	16.13	16.14	15.56	15.51	14.0	13.03	9.82	9.45
MG 2	7.07	6.41	5.46	5.27	6.01	6.43	7.15	7.12
MG 3	15.14	15.31	15.46	15.75	16.87	17.34	17.93	18.21
MG 4	4.85	4.91	4.93	5.01	5.71	5.62	5.91	5.64
MG 5	9.06	9.01	9.31	8.41	8.06	7.51	7.33	6.36

Table 9.3 Adjustable load.

Microgrid	Load	Type	Min.-max. capacity (MW)	Required energy (MWh)	Required start-end time (h)	Min. up time (h)
MG 1	L1	S	0–0.4	1.6	11–15	1
	L2	S	0–0.4	1.6	15–19	1
	L3	S	0.02–0.8	2.4	16–18	1
	L4	S	0.02–0.8	2.4	14–22	1
	L5	C	1.8–2	47	1–24	24
MG 2	L1	S	0–0.4	1.6	16–12	1
	L2	S	0.02–0.8	2.4	15–23	1
	L3	C	1.8–2	47	1–24	24
MG 4	L1	S	0–0.4	1.6	1–5	1
	L2	S	0.02–0.8	2.4	6–14	1

S: Shiftable, C: Curtailable.

Table 9.4 Dispatchable units.

Microgrid	Unit	Cost coefficient ($/MWh)	Min.–max. capacity (MW)	Min. up/down time (h)	Ramp up/down rate (MW/h)
MG 1	G1	27.7	1–5	3	2.5
	G2	39.1	1–5	3	2.5
	G3	61.3	0.8–3	1	3
	G4	65.6	0.8–3	1	3
MG 2	G1	30.9	1–2	4	1
	G2	45.7	0.5–2	4	2
	G3	73.5	0.5–1	2	1
	G4	78.4	1–3	3	1.5
MG 3	G1	25.6	1.5–3	6	1
	G2	28.3	1.5–4	6	2.5
	G3	63.9	0.5–3	4	2
	G4	67.1	0.5–3	4	2.5
	G5	89.2	1–6	3	3
MG 4	G1	23.3	0.5–1	2	1
	G2	51.6	0.5–3	4	2
	G3	69.2	1–3	5	1.5
MG 5	G1	23.6	0.5–1	4	1
	G2	35.4	1–2	3	2
	G3	45.7	0.5–2	4	2
	G4	63.1	1–3	3	1.5

Table 9.5 Distributed energy storages.

Microgrid	Storage	Capacity (MWh)	Min.–max. charging/discharging power (MW)	Min. charging/discharging time (h)
MG 1	DES	10	0.4–2	5
MG 2	DES	5	0.2–1	4
MG 3	DES	6	1–2	3
MG 4	DES	8	0.5–2	4
MG 5	DES	4	0.5–1	4

Table 9.6 Hourly market price ($/MWh).

Time (h)	**1**	**2**	**3**	**4**	**5**	**6**	**7**	**8**
Price ($/MWh)	15.03	10.97	13.51	15.36	18.51	21.80	17.30	22.83
Time (h)	9	10	11	12	13	14	15	16
Price ($/MWh)	21.84	27.09	37.06	68.95	65.79	66.57	65.44	79.79
Time (h)	17	18	19	20	21	22	23	24
Price ($/MWh)	115.45	110.28	96.05	90.53	77.38	70.95	59.42	56.68

Table 9.7 Super-microgrid formation in a 24-hour horizon.

	Hours (1–24)																							
MG1	1	1	1	1	1	1	1	1	1	1	1	1	1	1	1	1	1	1	1	1	1	1	1	1
MG2	2	2	2	2	2	2	1	2	2	2	2	2	2	2	2	1	2	2	2	2	2	2	1	2
MG3	1	3	3	3	2	2	1	3	3	3	2	3	3	1	3	1	1	3	3	3	1	3	2	3
MG4	2	1	2	4	3	2	1	3	3	3	3	1	4	3	4	2	1	3	3	2	1	3	3	3
MG5	1	2	3	5	4	3	5	1	4	3	4	2	5	1	2	1	2	2	3	2	1	1	4	1

operation cost. Each super-microgrid is formed so that the net spinning reserve of that specific super-microgrid is minimized. In other words, among all possible combinations to form a super-microgrid, those combinations that provide the least net spinning reserve during the islanded operation are the desired ones.

Table 9.7 illustrates how the super-microgrids are formed in a 24-hour scheduling horizon to minimize the system-aggregated operation cost in the grid-connected mode and the costs of power deficiency and spinning reserve in the islanded mode. The numbers utilized in the table represent each microgrid belonging to a super-microgrid in each hour. For instance, at $t = 1$, two super-microgrids are formed; microgrids 1, 3, and 5 are in one super-microgrid, while microgrids 2 and 4 have formed another super-microgrid. In hours such as $t = 4$, 7, and 13, the microgrids do not have an adequate spinning reserve to offer to adjacent microgrids, so super-microgrids are not formed.

Spinning reserve and power deficiency of microgrids at hours 17–20 are tabulated in Table 9.8 as examples. At $t = 17$, two super-microgrids are formed in a way that would make the net spinning reserve of the entire system zero. In this hour, microgrids 1 and 4 have 0.32 and 0.30 MW spinning reserves, respectively, which together cover the 0.62 MW power deficiency of microgrid 3. In addition,

Table 9.8 Spinning reserve and power deficiency of microgrids in four hours as a sample (MW).

Time (h)	17	18	19	20
MG 1	0.32	0	0	0
MG 2	1.48	1.60	0	1.50
MG 3	−0.62	−0.74	0.42	0
MG 4	0.30	0.44	0.93	0.13
MG 5	−1.48	−1.60	−1.35	−1.63

the 1.48 MW power deficiency of microgrid 5 is supplied by the 1.48 MW spinning reserve of microgrid 2. At $t = 18$, microgrids 3 and 4 form a super-microgrid; however, the spinning reserve of microgrid 4 is not adequate to supply the power deficiency of microgrid 3, and consequently, this super-microgrid experiences 0.3 MW load curtailment. On the other hand, the other formed super-microgrid in this hour does not experience any load curtailment as the spinning reserve of microgrid 2 supports the power deficiency of microgrid 5.

Figure 9.2 demonstrates the load curtailment of microgrids before and after the super-microgrid formations. As this figure shows, before super-microgrid formations, the system experiences load curtailment between hours 14 and 21, while after super-microgrid formations, the system faces load curtailment only at hour 18. The total load curtailment before super-microgrid formations is 10.08 MWh,

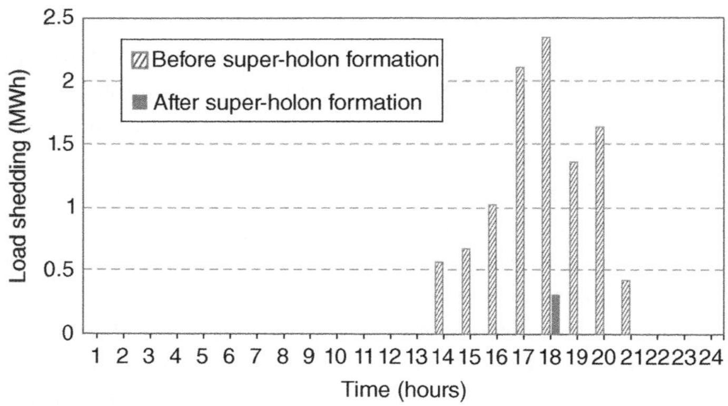

Figure 9.2 Load curtailment of integrated microgrids, before and after super-microgrid formation.

whereas this number drops to 0.3 MWh after super-microgrid formation. Nevertheless, this significant decrease in load curtailment after the supper-microgrid formation, which improves the system resilience, advocates for the effectiveness of the proposed model.

9.5 Summary

A spinning reserve-based optimal scheduling model of integrated microgrids in adaptive distribution grids for improving resilience against disasters was proposed in this chapter. The proposed model was focused on solving the common convergence issues with the existing models in the literature by moving away from a power exchange-focused modeling and adopting a spinning reserve-based approach. As the spinning reserve provision is costly, identifying the optimal super-microgrid combinations based on the minimum net spinning reserve was the main goal of the proposed model. Nevertheless, the configuration of connecting microgrids and forming super-microgrids was optimally determined via a spinning reserve-based integrated microgrids scheduling model, which subsequently improved microgrids' economic benefits and significantly enhanced the entire system's reliability and resilience.

References

1 Microgrid Exchange Group, DOE. *Microgrid Workshop Report*. Aug. 2011. Accessed: Jan 21, 2023. [Online]. Available: http://energy.gov/oe/downloads/microgrid-workshop-report-august-2011 [Accessed: January 21, 2023]

2 N. Hatziargyriou, H. Asano, M. R. Iravani, and C. Marnay, "Microgrids: An overview of ongoing research, development and demonstration projects," *IEEE Power Energy Mag.*, vol. 5, no. 4, pp. 78–94, Jul./Aug. 2007.

3 M. Shahidehpour, "Role of smart microgrid in a perfect power system," in *Proc. IEEE Power Energy Soc. Gen. Meeting*, Minneapolis, MN, USA, 2010.

4 B. Kroposki, R. Lasseter, T. Ise, S. Morozumi, S. Papathanassiou, and N. Hatziargyriou, "Making microgrids work," *IEEE Power Energy Mag.*, vol. 6, no. 3, pp. 40–53, May 2008.

5 A. Khodaei, "Resiliency-oriented microgrid optimal scheduling," *IEEE Trans. Smart Grid*, vol. 5, no. 4, pp. 1584–1591, Jul. 2014.

6 F. Katiraei and M. R. Iravani, "Power management strategies for a microgrid with multiple distributed generation units," *IEEE Trans. Power Syst.*, vol. 21, no. 4, pp. 1821–1831, Nov. 2006.

7 I. Bae and J. Kim, "Reliability evaluation of customers in a microgrid," *IEEE Trans. Power Syst.*, vol. 23, no. 3, pp. 1416–1422, Aug. 2008.

8 A. Khodaei and M. Shahidehpour, "Microgrid-based co-optimization of generation and transmission planning in power systems," *IEEE Trans. Power Syst.*, vol. 28, no. 2, pp. 1582–1590, May 2013.

9 S. Kennedy and M. Marden, "Reliability of islanded microgrids with stochastic generation and prioritized load," in *Proc. IEEE Bucharest Powertech*, Romania, Bucharest, Jun. 2009, pp. 1–7.

10 A. G. Tsikalakis and N. D. Hatziargyriou, "Centralized control for optimizing microgrids," *IEEE Trans. Energy Convers.*, vol. 23, no. 1, pp. 241–248, Mar. 2008.

11 S. Bahramirad, A. Khodaei, J. Svachula, and J. R. Aguero, "Building resilient integrated grids: One neighborhood at a time," *IEEE Electrif. Mag.*, vol. 3, no. 1, pp. 48–55, Mar. 2015.

12 C. Hou, X. Hu, and D. Hui, "Hierarchical control techniques applied in micro-grid," in *Proc. IEEE Conf. Power Syst. Tech. (POWERCON)*, Hangzhou, China, Oct. 2010, pp. 1–5.

13 A. Khodaei, "Microgrid optimal scheduling with multi-period islanding constraints," *IEEE Trans. Power Syst.*, vol. 29, no. 3, pp. 1383–1392, May 2014.

14 M. Shahidehpour and J. Clair, "A functional microgrid for enhancing reliability, sustainability, and energy efficiency," *Elect. J.*, vol. 25, no. 8, pp. 21–28, Oct. 2012.

15 A. Flueck and Z. Li, "Destination perfection," *IEEE Power Energy Mag.*, vol. 6, no. 6, pp. 36–47, Nov./Dec. 2008.

16 S. Bahramirad, W. Reder, and A. Khodaei, "Reliability-constrained optimal sizing of energy storage system in a microgrid," *IEEE Trans. Smart Grid*, vol. 3, no. 4, pp. 2056–2062, Dec. 2012.

17 A. Albaker and A. Khodaei, "Elevating prosumers to provisional microgrids," in *Proc. IEEE Power Energy Soc. Gen. Meeting*, Chicago, IL, USA, 2017.

18 M. Sandoval and S. Grijalva, "Future grid business model innovation: A prosumer-based cost-benefit framework for valuation of distributed energy resources," *IEEE PES Innov. Smart Grid Techn. Latin America (ISGT LATAM)*, Montevideo, Uruguay, 2015, pp. 450–455.

19 A. Khodaei, "Provisional microgrids," *IEEE Trans. Smart Grid*, vol. 6, no. 3, pp. 1107–1115, May 2015.

20 A. Albaker and A. Khodaei, "Valuation of microgrid unused capacity in islanded operation," in *Proc. IEEE North American Power Symp. (NAPS)*, Morgantown, WV, USA, 2017.

21 L. Che, X. Zhang, M. Shahidehpour, A. Alabdulwahab, and A. Abusorrah, "Optimal interconnection planning of community microgrids with renewable energy sources," *IEEE Trans. Smart Grid*, vol. 8, no. 3, p. 1054, May 2017.

22 Md. Asaduz-Zaman and A. H. Chowdhury, "Optimum economic dispatch of interconnected microgrid with energy storage system," in *Int. Conf. Electr. Eng. Inf. Commun. Technol. (ICEEICT)*, Dhaka, Bangladesh, May 2015, pp. 1–6.

23 W. Xi, Q. Xiaoyan, J. Runzhou, L. Dan, W. Gang, and L. Qian, "Economic operation of multi-microgrids containing energy storage system," in *Int. Conf. Power Syst. Technol. (POWERCON)*, Chengdu, China, Oct. 2014, pp. 1712–1716.

24 A. Albaker and A. Khodaei, "Communicative scheduling of integrated microgrids," in *Proc. IEEE Power Energy Soc. T&D*, Denver, CO, USA, 2018.

25 N. Nikmehr and S.N. Ravadanegh, "Optimal power dispatch of multi-microgrids at future smart distribution grids," *IEEE Trans. Smart Grid*, vol. 6, no. 4, pp. 1648–1657, Jul. 2015.

26 A.K. Marvasti, Y. Fu, S. DorMohammadi, and M. Rais-Rohani, "Optimal operation of active distribution grids: a system of systems framework," *IEEE Trans. Smart Grid*, vol. 5, no. 3, pp. 1228–1237, May 2014.

27 S. Howell, Y. Rezgui, J. L. Hippolyte, B. Jayan, and H. Li, "Towards the next generation of smart grids: Semantic and holonic multi-agent management of distributed energy resources," *Renew. Sustain. Energy Rev.*, vol. 77, pp. 193–214, 2017.

28 A. Ferreira, A. Ferreira, O. Cardin, and P. Leitão, "Extension of holonic paradigm to smart grids," *IFAC-PapersOnLine*, vol. 48, no. 3, pp. 1099–1104, 2015.

29 A. Koestler, *The Ghost in the Machine*. New York, NY, USA: Penguin Group, 1990.

30 A. Albaker and A. Khodaei, "Optimal scheduling of integrated microgrids in holonic distribution grids," in *Proc. IEEE North American Power Symp. (NAPS)*, Fargo, ND, USA, 2018.

31 A. Albaker and A. Khodaei, "Spinning reserve based topology control in holonic distribution grids," in *CIGRE Grid Future Symp.*, Reston, VA, USA, 2018.

32 E. Negeri, N. Baken, and M. Popov, "Holonic architecture of the smart grid," *Smart Grid Renew. Energy*, vol. 4, no. 2, p. 202, 2013.

33 S. Frey, A. Diaconescu, D. Menga, and I. Demeure, "A holonic control architecture for a heterogeneous multi-objective smart micro-grid," in *IEEE Self-Adaptive Self-Org. Syst. (SASO)*, Sept. 2013.

34 A. Pahwa *et al.*, "Goal-based holonic multiagent system for operation of power distribution systems," *IEEE Trans. Smart Grid*, vol. 6, no. 5, pp. 2510–2518, Sep. 2015.

35 A. Pahwa *et al.*, "Holonic multi-agent control of power distribution systems of the future," in *Grid Future Symp. (CIGRE)*, 2012.

36 G. Florea, O. Chenaru, R. Dobrescu, and D. Popescu, "A fractal model for power smart grids," in *IEEE 20th Int. Conf. Control Syst. Comput. Sci. (CSCS)*, May 2015, pp. 572–577.

37 F. Ounnar, A. Naamane, P. Pujo, and N. K. M'Sirdi, "Intelligent control of renewable holonic energy systems," *Energy Procedia*, vol. 42, pp. 465–472, 2013.

38 Z. Xiaohui, G. Wenbo and Z. Jiaqing, "Decentralized economic dispatching of multi-micro grid considering wind power and photovoltaic output uncertainty," *IEEE Access*, vol. 9, pp. 104093–104103, 2021, doi: https://doi.org/10.1109/ACCESS.2021.3099401.

39 M. Thirumalai, R. Hariharan, T. Yuvaraj, and N. Prabaharan. "Optimizing distribution system resilience in extreme weather using prosumer-centric microgrids with integrated distributed energy resources and battery electric vehicles," *Sustainability*, vol. 16, no. 6, p. 2379, 2024.

40 N. Nikmehr, P. Zhang, H. Zheng, T. C. Wei, G. He, and Y. A. Shamash, "Quantum annealing-infused microgrids formation: distribution system restoration and resilience enhancement," *IEEE Trans. Power Syst.*, doi: https://doi.org/10.1109/TPWRS.2024.3399122.

41 ILOG CPLEX, "ILOG CPLEX homepage." 2009. Accessed: Dec 2009. [Online]. Available: http://www.ilog.com

10

The Economics of Post-Disaster Restoration

10.1 Principles of Power System Restoration

Climate shocks and natural disasters around the world resulted in significant economic, social, and physical disruptions and caused considerable inconvenience for residents living in disaster areas due to the loss of electricity, water, and communications. A resilient power system encounters minimum possible outages and will quickly return to its normal operating state [1]. "After a storm comes a calm" does not hold for power infrastructure. The increasing intensity and frequency of natural disasters due to climate change call for devising efficient strategies for pre- and post-disaster power infrastructure management. Mitigating the aftermath of disasters by improving the power grid resilience, as one of the critical lifeline systems, is of utmost importance for utility companies and governments.

Over half of a century after the publication of one of the earliest studies on efficient response to natural disasters, motivated by Hurricane Carla that slammed into the Gulf Coast and moved onward into the United States and Canada [2], the issue of efficient response to disasters still seems to remain in its immature stage. Grid operators commonly face two challenges in response to the damaged power grid due to disasters: first, in the planning stage to design more reliable networks, and second, in the operation stage, in which they attempt to manage the restoration process in an efficient way [3]. In this chapter, the focus will be on the operation stage. However, the literature in both the planning and operation stages is explored to provide broader insight for the interested readers.

10.1.1 Planning Stage

Various studies have been proposed in the literature in the context of emergency planning for power grids. In [4], the research problems and models for substations and/or distribution feeders planning under normal and emergency conditions were reviewed and discussed. A case study on hurricane planning and rebuilding the electrical infrastructure along the Gulf Coast for Hurricane Katrina was presented in [5]. A risk assessment method for infrastructure technology planning to improve the power supply resiliency to natural disasters was proposed in [6]. Reduced cost and power supply availability were considered two fundamental decision factors in their disaster planning approach. In [7], a stochastic integer program was proposed to find the optimal schedule for inspection, damage evaluation, and repair of electric power grids in post-earthquake restoration to minimize the average time that each customer is without power. Bienstock and Mattia [8] proposed two models to solve power grid blackout problems using mixed-integer programming. The optimization problems relevant to preventing large-scale blackouts in transmission grids subject to stochastic damage scenarios were considered. The first model decides which transmission lines are to be expanded in capacity to guarantee that all power flows are within the desired limits after damage to transmission lines in different scenarios. The second model considers the dynamics of cascades to find an optimal reinforcement plan that can passively survive a potential cascade. A comprehensive survey of models and algorithms for emergency response logistics in distribution grids, including reliability planning with fault considerations and contingency planning models, were presented in [9, 10].

In the context of resource allocation for power grid restoration, [11] presented three mathematical programming models to locate the repair units and restore the transmission and distribution lines efficiently. The first model finds the optimal repair-unit dispatch tactical plan with a forecast of adverse weather conditions. The second model derives the optimal repair-unit location for a short-term strategic plan under normal weather conditions. The third model finds the optimal number of repair units for a long-term strategic plan. In [12], a mixed-integer programming model and a general column-generation approach for inventory decision-making of power system components throughout a populated area to maximize the amount of power served after disaster restoration was proposed. In [13], the service restoration considering the restrictions on emergency response logistics with the objective of minimizing the customers interruption cost was studied. The reconfiguration and the resource dispatching issues were considered systematically to derive the optimal time sequence for every step of the restoration plan. In [14], a decision-making model was proposed to manage the required resources for economic power restoration. The optimal number of depots, the optimal location of depots, and the optimal number of repair crews were determined in their model to minimize the transportation cost associated with restoration operations. In [15], a decision support tool for improving information used by electric utilities for managing the restoration of distribution grid components damaged due to large-scale storms was described. The circuit layout, the placement of protective and switching devices, and the location of customers were considered to allocate the crew in a cost-effective manner.

10.1.2 Operating Stage

In the context of restoration, [3] studied the budgeted and the minimum weighted latency variants of the recovery problem of large-scale power outages due to a major disaster. The problems for general cases, as well as trees and bipartite networks as special cases, were studied. Authors in [16] formulated a mixed-integer program to model the recovery of the transmission networks damaged due to disasters. The model considers the repair crew constraints as well as the penalty cost of unserved loads to find the recovery schedule, which minimizes the cost of a power outage. In [17], a mixed-integer programming framework is used for modeling the optimal supply restoration of the faulty distribution grids. A two-step decomposition method was developed to derive the optimal configuration as well as the optimal switching sequence of the distribution grid. Authors in [18] studied three approaches for joint damage assessment and restoration of the power grids after natural disasters. The proposed approaches include an online stochastic combinatorial optimization algorithm that dynamically makes the restoration decisions once each potentially damaged site is visited. This two-stage method first evaluates the extent of the damage and then restores the system, and a hybrid algorithm of both approaches, which simultaneously performs the damage evaluation and system restoration tasks. The results indicate that the first approach can provide higher-quality solutions for joint damage assessment and recovery problems. Authors in [19] proposed a general multi-objective linear-integer spatial optimization model for arcs and nodes restoration of disrupted networked infrastructure after disaster. The proposed model addresses the tradeoff between maximizing the system power flow and minimizing the system cost. Studies in [20] offered an integrated network design and scheduling problem for restoring the interdependent civil infrastructure. The problem was formulated using integer programming and analyzed on a realistic data set of power infrastructure of Lower Manhattan in New York City and New Hanover County, North Carolina. The results indicate that the proposed model can be used for real-time as well as long-term restoration planning. [21] considered the last-mile restoration of power systems, i.e., how to schedule and allocate the routes to fleets of repair crews to recover the damaged power grid as quickly as possible. The power grid restoration and vehicle routing were decoupled to improve the computational efficiency of the model. The results show that the proposed model outperforms the models practiced in the field regarding solution quality and scalability. This work was extended in [22] by applying a randomized adaptive vehicle decomposition technique to improve the model's scalability for large-scale disaster restoration of power grids with more than 24,000 components. In another work, [23] presented a scalable approach for restoring the interdependent gas and power infrastructure. Mixed-integer programming was used to obtain a minimum restoration set and optimal restoration ordering. Randomized adaptive decomposition was applied to improve the solution quality and computational efficiency. Ref. [24] introduced a multi-stage dynamic recovery model aimed at enhancing the resilience of distribution networks in the aftermath of disasters. The proposed model integrates energy systems and transportation networks. The first stage involves rapidly increasing the output of gas turbines and reconfiguring distribution into microgrids to maximize load restoration. The second stage deploys mobile energy storage systems to address power shortages and further restore loads. The third stage involves adjusting recovery measures for economically optimized operations. Ref. [25] proposed a model for coordinating repair and dispatch resources to restore power in distribution systems after extreme weather events. It aims to address uncertainties in the output of renewable energy resources and load demand by formulating a two-stage optimization model. The proposed model optimizes the dispatch and operation of repair crews, mobile power sources, energy storage systems, and network reconfiguration to maximize the restored load. It involves re-optimizing the coordination model to handle unexpected changes in network topology and distributed resources. Authors in [26] presented a model for optimal planning of coupled photovoltaics-energy storage-charging stations (PV-ES-CS) within hybrid AC/DC distribution networks, aiming at balancing economic efficiency and resilience to extreme events. They proposed a bi-level optimization model, utilizing a Nash equilibrium-based payment function to balance these objectives at the upper level, and employing a genetic algorithm to simulate scenarios and obtain optimal results at the lower level. Key findings indicate that integrating PV-ES-CS can enhance network resilience and economic performance compared to a single-objective model. In our previous work [27, 28], we proposed reactive and proactive restoration strategies for power system infrastructure considering natural disaster effects. In [29], we proposed a dynamic maintenance model for power systems subject to failure due to natural disasters.

In this chapter, we present a generalized post-disaster restoration scheme for power grids by simultaneously incorporating the physics of the system and the economics of disaster in the restoration process. Intuitively, ignoring the economics of restoration and underlying constraints results in a suboptimal or infeasible restoration plan. The unit commitment and optimal power flow problems, along with the resource cost and the opportunity cost of load interruption defined by the value of lost load (VOLL), are considered economic measures in the model. These economic considerations impose additional constraints and redefine the restoration cost function. The restoration problem for a DC power flow model is formulated as a mixed-integer program. By a mixed-integer programming approach, different combinations of restoration schedules and the operational configuration of the grid are searched in a large solution space to find a cost-effective restoration plan. The model intends to minimize the customer load interruption, restoration operation, and electricity generation costs without violating the system's physics. The output of the proposed model includes the post-disaster restoration schedule, generation unit commitment states, power dispatch, and transmission grid configuration. Therefore, the proposed decision-making model not only determines the restoration schedule but also provides a practical and cost-effective operational configuration for major power grid components during the restoration time horizon.

10.2 Economic Variables in System Restoration

Consider that the main components of a power infrastructure system, i.e., generation units, transmission lines, buses (substations), and downstream distribution lines, are subject to damage due to climate shocks and natural disasters. In this post-disaster restoration model, we present a decision-making tool for a typical vertically integrated utility company to schedule the repair operations of its transmission lines and buses in coordination with the operations of its generation units in a cost-effective manner. Independent System Operator (ISO) can also use this model in a restructured power market in coordination with associated electric utilities.

Once the disaster strikes, the utility company conducts a damage assessment by an aerial survey of the power network (by using helicopters, drones, or satellite technologies) in affected areas, as well as a ground check by inspectors (if the roads are not affected) [30]. Damage assessment determines whether a facility/component has been damaged at all, and if damaged, estimates the expected Time to Repair (TTR) of that component. Each bus, along with its downstream distribution lines, is consolidated and considered a single component, resulting in a consolidated TTR. While the generation units are part of the vertically integrated utility company, it is assumed that each generation unit is responsible for the repair operations of its damaged facilities. Therefore, each of the damaged generation units submits its repair operations schedule to the operations coordinator of the vertically integrated utility company to be used as input for the restoration scheduling of transmission lines and buses in a coordinated manner. We consider two states for each component: *damaged* if the component has encountered major damages, and thus it is offline and needs to be restored; and *functional* if it has not been damaged at all, or minor damages have occurred, and the component can continue its functionality.

After determining each component's initial damaged or functional state, the restoration resources need to be allocated to repair the damaged components. Without loss of generality and for proof of concept, we consider the repair crews as the only limited resource to be allocated to the damaged components. At the same time, it is assumed that plenty of spare parts and equipment for restoration are available. The resource allocation, however, is subject to the criticality of the load to be restored as well as costs associated with seizing the resources at each particular time and location. In this regard, the objective of the problem is defined as minimizing the customer interruption cost, plus the restoration resource cost and power generation cost. The interruption cost is the amount of the interrupted load multiplied by VOLL. VOLL is an important measure in the electricity market's micro- and macroeconomics which represents the willingness of customers to pay for their electricity service to avoid curtailment. Valuation of VOLL, usually measured in dollars per MWh, can either be based on the marginal value of the next unit of interrupted electricity load or the average value of the interrupted load. However, it varies depending on the type of usage and outage [31]. Using the lens of economics, the load interruption cost is considered an opportunity cost.

10.3 A Generic Economic Model for Post-Disaster Restoration

10.3.1 Objective Function

The objective is to minimize the customer load interruption cost, the restoration operation cost, and the power generation cost as follows:

$$\min_{u,v,LI,P,SU.SD} \sum_t \sum_b VOLL_{bt}\, LI_{bt} + \sum_t \sum_b C_{bt} R_b u_{bt} + \sum_t \sum_l C_{lt} R_l u_{lt} + \sum_t \sum_i C^g_{it} P_{it} + SU_{it} + SD_{it} \tag{10.1}$$

where the first term represents the total opportunity cost of load interruption over the restoration planning horizon in which LI_{bt} represents the load interruption at bus b at time t. The second term represents the cost of resources allocated to the buses (and their downstream distribution lines), where C_{bt} is the hourly crew cost per person to repair bus b, and R_b represents the crew size to repair bus b at time t. The third term is the cost of resources allocated to transmission lines, where C_{lt} represents the hourly crew cost per person to repair line l, and R_l is the crew size to repair line l at time. The fourth term indicates the generation cost of all the units feeding the system, including the fuel costs and the startup/shut down costs where C^g_{it} is the generation cost of unit i at time t, SU_{it} is the startup cost variable of unit i at time t, SD_{it} is the shutdown cost variable of unit i at time t, and P_{it} is the real power generation of unit i at time t. The resource cost is defined as the summation of the product of the number of allocated resource(s) to each component and the cost of each unit of resource seized by that component. The binary decision variables u_{bt} and v_{lt} indicate whether or not a unit of crew resources is allocated at each particular time period to each particular bus and line, respectively (1 if allocated, otherwise 0).

10.3.2 Constraints

10.3.2.1 Damage State Modeling

A mechanism is required in the model to allocate the resources only to the damaged components. This mechanism also needs to stop the allocation of resources and switch the damaged state of the components into a functional state when the resources are seized for a predetermined duration of TTR. This mechanism for generation units is modeled as follows:

$$y_{it} = 0, \quad \textit{if } t \leq TTR_i; \quad \textit{otherwise } y_{it} = 1, \quad \forall i, \forall t \tag{10.2}$$

However, since the *if-then* constraint is not allowed in linear programming, constraint (10.2) is decomposed and rewritten as follows:

$$t - My_{it} \leq TTR_i, \quad \forall i, \forall t \tag{10.3}$$

$$My_{it} \leq TTR_i, \quad \forall i, \forall t = 0, 1, \dots, TTR_i \tag{10.4}$$

where

binary variable y_{it} represents the damaged or functional state of generation unit i at time t

M is a large positive number

Binary variable y_{it} is equal to 0 if generation unit i is damaged due to the disaster and has not been restored until time t; otherwise, it is equal to 1. The time it takes from the beginning of the planning horizon for a damaged generation unit to be repaired and brought back to the system is represented by TTR_i. If the component has not encountered any damage, the time to repair TTR_i is set to 0.

Constraints (10.5) and (10.6) present the relationship between binary state variables w_{lt} and z_{bt} with their corresponding repair decision variables v_{lt} and u_{bt}, respectively:

$$0 \leq w_{l(t+1)} - \left(\sum_{k=1}^{t} v_{lk} - TTR_l + 0.5 \right) / M \leq 1 \quad \forall l, \forall t \tag{10.5}$$

$$0 \leq z_{b(t+1)} - \left(\sum_{k=1}^{t} u_{bk} - TTRb + 0.5 \right) / M \leq 1 \quad \forall b, \forall t \tag{10.6}$$

where binary variables z_{bt} and w_{lt} are the damaged or functional state of bus b at time t, and transmission line l at time t, respectively. If the transmission line l at time t is on damaged state, the binary variable w_{lt} becomes 0. Once it is repaired, the value of w_{lt} becomes 1 and remains the same up to the end of the restoration planning horizon. Binary variable v_{lt} is the decision variable for repair of line l. When line l is under repair at time t, decision variable v_{lt} takes the value of 1, otherwise 0. Similarly, z_{bt} is equal to 0 when bus b at time t is in a *damaged* state. Once it is repaired, the value of z_{bt} becomes 1 and remains the same up to the end of the planning horizon. Binary variable u_{bt} is the decision variable for the repair of bus b, which takes the value of 1 when the bus b is under repair, otherwise, it is equal to 0.

Constraints (10.7) and (10.8) further guarantee that enough time and resources are allocated to restore each damaged component:

$$\sum_{k=t}^{t+TTR_l-1} v_{lk} \geq TTR_l \left(v_{lt} - v_{l(t-1)} \right), \quad \forall l, \forall t \tag{10.7}$$

$$\sum_{k=t}^{t+TTR_b-1} u_{bk} \geq TTR_b \left(u_{bt} - u_{b(t-1)} \right), \quad \forall b, \forall t \tag{10.8}$$

We assume that once the restoration operation on a particular component is started, it should continue at least for the estimated TTR duration. The addition of these constraints eliminates partial repair of damaged components. Finally, constraints (10.9)–(10.11) narrow down the solution space and eliminate unnecessary allocation of resources to repaired components:

$$y_{i(t+1)} \geq y_{it}, \quad \forall i, \forall t \tag{10.9}$$

$$w_{l(t+1)} \geq w_{lt}, \quad \forall l, \forall t \tag{10.10}$$

$$z_{b(t+1)} \geq y_{bt}, \quad \forall b, \forall t \tag{10.11}$$

10.3.2.2 Resource Constraints

The objective function of the model is constrained by limited restoration resources, as follows:

$$\sum_{l} R_l v_{lt} + \sum_{b} R_b u_{bt} \leq R_t^{max}, \quad \forall t \tag{10.12}$$

Constraint (10.12) represents the maximum number of resources that can be allocated to the entire system in each unit of time, where R_t^{max} represents the number of available repair crew at time t.

10.3.2.3 Load Balance Constraints

The physics of the grid imposes another constraint, i.e., the bus load balance constraint, to the objective function of the post-disaster model, as follows:

$$\sum_{i \in N_b} P_{it} + \sum_{l \in N_b} PL_{lt} + LI_{bt} = D_{bt}, \quad \forall b, \forall t \tag{10.13}$$

where

N_b is the set of components connected to bus b
PL_{it} is the power flow of line l at time t.

The bus load balance constraint (10.13) ensures that the injected power to a bus from connected transmission lines and generation units is fully supplying the bus load at time t which is D_{bt}; however, if the injected power is not sufficient, the load is interrupted (modeled by a load interruption variable LI_{bt}). The load interruption variable is non-negative and smaller than the load at its associated bus.

10.3.2.4 Real Power Generation Constraints

The real power generation in each unit i is bounded by its damage state, unit commitment state (I_{it} is the commitment state of generation unit i at time t), and minimum and maximum generation capacity (represented by P_i^{min} and P_i^{max}, respectively), as follows:

$$P_i^{min} y_{it} I_{it} \leq P_{it} \leq P_i^{max} y_{it} I_{it}, \quad \forall i, \forall t \tag{10.14}$$

The real power generation constraint (10.14) is nonlinear. We linearize this constraint by defining a new variable $n_{it} = y_{it} I_{it}$ and using the following set of equations:

$$P_i^{min} n_{it} \leq P_{it} \leq P_i^{max} n_{it}, \quad \forall i, \forall t \tag{10.15}$$

$$n_{it} - y_{it} \leq 0, \quad \forall i, \forall t \tag{10.16}$$

$$n_{it} - I_{it} \leq 0, \quad \forall i, \forall t \tag{10.17}$$

$$-n_{it} + y_{it} + I_{it} \leq 1, \quad \forall i, \forall t \tag{10.18}$$

$$n_{it} \geq 0. \quad \forall i, \forall t \tag{10.19}$$

It is important to note that if a generation unit is not in the *functional* state, it cannot be committed for a generation. Therefore, the coupling constraint of unit commitment and damage state holds all the time, i.e.,

$$I_{it} \leq y_{it}, \quad \forall i, \forall t \tag{10.20}$$

The damage state of bus(es) connected to each generation unit constrains the real power generation, as follows:

$$-M\sum_{b}\alpha_{ib}z_{bt} \leq P_{it} \leq M\sum_{b}\alpha_{ib}z_{bt}, \quad \forall i, \forall t \tag{10.21}$$

where α_{ib} is the element of the generation-bus incidence matrix that takes the value of 1 if generation unit i is connected to bus b; otherwise, 0. If a connected bus to a generation unit is damaged, the associated generation unit becomes offline.

10.3.2.5 Power Flow Constraints

Incorporating the physics of the grid, the damage state of each transmission line along with associated bus(es) and their impacts on power flow are modeled as follows:

$$-PL_l^{max}w_{lt} \leq PL_{lt} \leq PL_l^{max}w_{lt}, \quad \forall l, \forall t \tag{10.22}$$

$$-M\sum_{b}\beta_{lb}^{from}z_{bt} \leq PL_{lt} \leq M\sum_{b}\beta_{lb}^{from}z_{bt}, \quad \forall l, \forall t \tag{10.23}$$

$$-M\sum_{b}\left|\beta_{lb}^{to}\right|z_{bt} \leq PL_{lt} \leq M\sum_{b}\left|\beta_{lb}^{to}\right|z_{bt}, \quad \forall l, \forall t \tag{10.24}$$

$$\begin{aligned} &-M(1-w_{lt}) - M\left(1-\sum_{b}|\beta_{lb}|z_{bt}\right) \\ &\leq PL_{lt} - \frac{\sum_b \beta_{lb}\delta_{bt}}{x_l} \leq M(1-w_{lt}) + M\left(1-\sum_{b}|\beta_{lb}|z_{bt}\right), \quad \forall l, \forall b, \forall t \end{aligned} \tag{10.25}$$

where

β_{lb} is the element of line l and bus b in line-bus incidence matrix
β_{lb}^{from} represents the positive elements of the bus-line incidence matrix
β_{lb}^{to} represents the negative elements of the bus-line incidence matrix

As shown in Eq. (10.22), if the line l at time t is in the *functional* state, the line can transfer power but is limited to its power flow capacity. However, if the line is damaged, the power flow will be equal to 0.

In addition, as long as any of the buses connected to each particular transmission line is in a *damaged* state, the power cannot flow in that particular line, as shown in Eqs. (10.23)–(10.25). Note that δ_{bt} is the bus voltage angle.

10.3.2.6 Startup and Shutdown Cost Constraints

The startup and shutdown costs have been defined in the objective function as positive variables to avoid using additional binary state variables to improve the computational efficiency of the program. Based on [32, 33], the startup and shutdown cost variables are bound to constraints (10.26)–(10.29), as follows:

$$SU_{it} \geq C_{i\tau}^{su}\left(I_{it} - \sum_{k=1}^{\tau} I_{i(t-k)}\right), \quad \forall i, \forall t, \forall \tau = 1, \ldots, ND_i \tag{10.26}$$

$$SU_{it} \geq 0, \quad \forall i, \forall t \tag{10.27}$$

$$SD_{it} \geq C_{it}^{sd}\left(I_{i(t-1)} - I_{it}\right) \quad \forall i, \forall t \tag{10.28}$$

$$SD_{it} \geq 0, \quad \forall i, \forall t \tag{10.29}$$

where

ND_i is the number of time intervals of the startup cost function for generation unit i
$C_{i\tau}^{su}$ is the startup cost parameter of unit i at time t
C_{it}^{sd} is the shutdown cost parameter of unit i at time t.

10.3.2.7 Full Restoration Constraint

There is a possibility that the system load will be fully recovered, while some generation units, transmission lines, or buses still have not been repaired. The reason is that the functional generation units, transmission lines, and buses temporarily compensate for the outage of redundant damaged components. Therefore, due to the economic dynamics of the system, the restoration process can be terminated by partial restoration. However, due to potential load increments, the system may not be able to fully and economically supply loads beyond the restoration horizon. To ensure that all damaged components are repaired by the end of the restoration horizon, the following constraint is added:

$$\sum_i y_{i(NT)} + \sum_b z_{b(NT)} + \sum_l w_{l(NT)} = NG + NB + NL \tag{10.30}$$

where

NT is the length of the restoration planning horizon
NG, NB, and NL are the number of generation units, buses, and transmission lines, respectively.

10.3.2.8 Ramp-Up and Ramp-Down Constraints

The mechanical and thermal inertia that should be overtaken to decrease (ramp-down) or increase (ramp-up) the real power generation of a thermal generation unit poses a variety of constraints to the system [34]. A computationally efficient mixed-integer linear formulation for the ramp-up rate and ramp-down rate was proposed by [32]. Our post-disaster restoration model is restricted to these constraints as follows:

$$P_{it} - P_{i(t-1)} \leq UR_i \cdot I_{i(t-1)} + UR_i^{su}\left(I_{it} - I_{i(t-1)}\right) + P_i^{max}(1 - I_{it}), \quad \forall i, \forall t \tag{10.31}$$

$$P_{it} \leq P_i^{max} \cdot I_{i(t+1)} + DR_i^{sd}\left(I_{it} - I_{i(t-1)}\right), \quad \forall i, \forall t \tag{10.32}$$

$$P_{i(t-1)} - P_{it} \leq DR_i \cdot I_{it} + DR_i^{sd}\left(I_{i(t-1)} - I_{it}\right) + P_i^{max}\left(1 - I_{i(t-1)}\right), \quad \forall i, \forall t \tag{10.33}$$

where the ramp-up rate is modeled in Eq. (10.31) in which UR_i is the ramp-up rate limit of unit i, the shutdown ramp rate is shown in Eq. (10.32) in which DR_i is the ramp-down rate limit of unit i, and the ramp-down rate is represented by Eq. (10.33) in the model.

10.3.2.9 Minimum Up-Time and Down-Time Constraints

In thermal generation units, the temperature change can only occur gradually. Thus, when a generation unit is operating, it cannot be decommitted immediately (minimum up time of generator i is UT_i), and once the unit is offline, it requires some time before it can be committed again (minimum down time of generator i is DT_i) [35]. A computationally efficient mixed-integer linear formulation for the minimum up-time and minimum-down time modeling was proposed by [32]. The post-disaster restoration model is restricted to these constraints as follows:

$$\sum_{t=1}^{G_i} \left(1 - I_{it}\right) = 0, \quad \forall i \tag{10.34}$$

$$\sum_{k=t}^{t+UT_i-1} I_{ik} \geq UT_i\left(I_{it} - I_{i(t-1)}\right), \quad \forall i, \forall t = G_{i+1}, \ldots, Nt - UT_i + 1 \tag{10.35}$$

$$\sum_{k=t}^{NT} \left(I_{ik} - \left(I_{it} - I_{i(t-1)}\right)\right) \geq 0, \quad \forall i, \forall t = NT - UT_i + 2, \ldots, Nt \tag{10.36}$$

$$\sum_{t=1}^{L_i} I_{it} = 0, \quad \forall i \tag{10.37}$$

$$\sum_{k=t}^{t+DT_i-1} \left(1 - I_{ik}\right) \geq DT_i\left(I_{i(t-1)} - I_{it}\right), \quad \forall i, \forall t = L_i + 1, \ldots, NT - DT_i + 1 \tag{10.38}$$

$$\sum_{k=t}^{NT} \left(1 - I_{ik} - \left(I_{i(t-1)} - I_{it}\right)\right) \geq 0, \quad \forall i, \forall t = NT - DT_i + 2, \ldots, NT \tag{10.39}$$

where

$G_i = min\left\{NT, \left(UT_i - U_i^0\right) I_i^0\right\}$ is the number of primary time periods that generation unit i is online

U_i^0 is the number of time periods up to the beginning of the planning horizon before generation unit i becomes online

I_i^0 is the primary commitment state of generation unit i

$L_i = min\left\{NT, \left(DT_i - S_i^0\right)\left(1 - I_i^0\right)\right\}$ is the number of primary time periods that generation unit i is offline

S_i^0 is the number of periods up to the beginning of planning horizon that generation unit i has been offline.

10.3.3 Decomposition Strategy

Benders decomposition has been widely used in the power system literature [36]. Benders decomposition for mixed-integer programming is an efficient strategy when the original problem is large-scale and difficult to solve. To employ the decomposition strategy for the proposed problem, we consider the continuous variable vector as $X = \left[LI_{bt}^T, P_{it}^T, PL_{it}^T, SU_{it}^T, SD_{it}^T\right]^T$ the binary variable vector as $Y = \left[u_{bt}^T, v_{lt}^T, y_{it}^T, z_{bt}^T, w_{lt}^T, I_{it}^T\right]^T$, the cost coefficient matrix of the integer variables in the objective function as C^T composed of $C_{bt}R_b$ and $C_{lt}R_l$, and the cost coefficient matrix of the continuous variables in the objective function as D^T composed of $VOLL_{bt}$ and **1**. We also assume that **A** and **B** represent the coefficient matrices of **X** and **Y** in the constraints, respectively. Finally, **H** is assumed to represent the right-hand-side matrix of constraints. Now, we can rewrite the proposed mixed-integer programming model in the following abstract form:

$$\min_{X,Y} C^T X + D^T Y$$

$$s.t. \; AX + BY \geq H \tag{10.40}$$

$$Y \in \{0,1\}, \quad X \geq 0 \tag{10.41}$$

The proposed problem is decomposed into an integer program (IP) master problem and a dual linear program (LP) subproblem. Considering **U** as the dual variable vector for the subproblem, the Benders decomposition algorithm for the proposed model is shown in Algorithm 1 [37]. In each iteration, after solving the master problem, the subproblem evaluates the obtained solution for feasibility. If the dual LP subproblem is found to be unbounded, the feasibility cuts are generated and added to the IP master problem; otherwise, the optimality cuts are generated to be added to the master problem to form a new objective function for the master problem (the constraints with θ represent the optimality cuts). The iterative process will continue until an acceptable relative gap (ε) between the current upper and lower bounds is obtained. The value of ε is considered to be 0.05 for this chapter.

Algorithm 10.1 Benders Decomposition for Mixed-Integer Program

{initialization}
Lower bound (LB):= $-\infty$, Upper bound (UB):= $+\infty$
while UB–LB > ε **do**
  {solve dual LP subproblem}
  $\max_U \left\{\sum_t \sum_b C_{bt}R_b\bar{u}_{bt} + \sum_t \sum_l C_{lt}R_l\bar{v}_{lt} + (H - B\bar{Y})^T U \mid A^T U \leq C, U \geq 0\right\}$
  if unbounded **then**
    Get unbounded ray $\bar{U}$
    Add cut $(H - BY)^T \bar{U} \leq 0$ to IP master problem
  else
    Get extreme point $\bar{U}$
    Add cut $\theta \geq \sum_t \sum_b C_{bt}R_b u_{bt} + \sum_t \sum_l C_{lt}R_l v_{lt} + (H - BY)^T \bar{U}$
      to master problem
    UB := min $\left\{\text{UB}, \sum_t \sum_b C_{bt}R_b\bar{u}_{bt} + \sum_t \sum_l C_{lt}R\bar{v}_{lt} + (H - B\bar{Y})^T \bar{U}\right\}$
  end if
  {solve IP master problem}
  $\min Y\{\theta \mid cuts, Y \in \{0, 1\}\}$
  LB := $\bar{\theta}$
end while

10.4 Case Study

The IEEE 118-bus system is used to analyze the post-disaster restoration model. The system has 118 buses, 54 generation units, 186 branches, and 91 load sides. The system setup is shown in Tables 10.1–10.3. Among damaged buses, B1, B2, B3, B4, B8, and B11 are load buses, feeding their downstream distribution lines, while B5 is not a load bus. From [31], the VOLL is assumed to be \$3.706/kWh for industrial loads, \$6.979/kWh for commercial loads, and \$0.110/kWh for residential areas. The VOLL for critical loads, e.g., medical centers and water treatment plants, can be misstated if an economics lens is adopted. However, due to the crucial importance of these critical loads, the VOLL in an *ad hoc* manner is considered to be \$10/kWh to impose higher priority on these areas. In this analysis, as shown in Table 10.1, bus B1 is considered commercial load, bus B11 industrial load, bus B4 critical load, and the rest of the load buses in the system are considered residential loads. The TTR in Tables 10.1 and 10.2 indicate the estimated duration

Table 10.1 Damaged buses and time to repairs.

Bus #	Time to repair (hours)	Load type
B1	24	Commercial
B2	11	Residential
B3	18	Residential
B4	15	Critical
B5	5	N/A
B8	4	Residential
B11	22	Industrial

Table 10.2 Damaged transmission lines and time to repairs.

Line #	Time to repair (hours)
L1	20
L2	18
L10	16
L14	10
L16	22

Table 10.3 Damaged generation units and time to repairs.

Unit #	Time to repair (hours)
G1	17
G2	12
G3	24
G4	8

of the repair for buses and transmission lines, respectively, while the TTR in Table 10.3 shows the time it takes from the beginning of the restoration planning horizon to repair and restore each damaged generation unit.

The repair crew is considered to be the only limited resource that is allocated to repair the damaged components. It is assumed that each damaged bus requires 12 repair crews/hour, while 18 repair crews/hour are required for each damaged transmission line. Although depending on skill levels, different crew costs can be considered, we assume that all repair crews have equal skill levels; hence, they are equally paid (we could think of it as a bundle of resources in which their average wage is used as an input into our model). The hourly wages for repair crews vary based on the working shift and type of repairs. For repairing the buses and downstream distribution lines, the average wages are assumed to be \$60/hour at shift 1 (8:00 A.M.–4:00 P.M.), \$70/hour at shift 2 (4:00 P.M.–12:00 A.M.), and. \$80/hour at shift 3 (12:00 A.M.–8:00 A.M.); for repairing the transmission lines, the average wages are assumed to be \$65/hour at shift 1, \$75/hour at shift 2, and \$85/hour at shift 3. Without loss of generality, it is assumed that all generation units incur identical generation, startup, and shutdown costs. From [38], the generation cost is considered to be \$0.3509/kWh. The shutdown cost is assumed to be \$250 per shutdown for each generation unit. The startup cost is assumed to be \$150 within the first hour after the last shutdown. For each additional hour (up to eight hours), an incremental cost of \$25 would be added to the startup cost. Restoration planning horizon starts at 8:00 AM. The length of the restoration planning horizon is set to be $NT = 120$ hours.

The following scenarios are adopted to analyze the model and the impacts of economic considerations in post-disaster restoration:

1) **Scenario I:** The problem is solved without considering the economics of disaster; only load interruption is minimized.
2) **Scenario II:** The problem is solved with consideration of the VOLL and repair cost. However, the generation cost is not considered.
3) **Scenario III:** The problem is solved with full economic consideration, i.e., VOLL, repair cost, and generation cost.
4) **Scenario IV:** The economic impact of the maximum number of available resources (repair crews) on restoration is analyzed for five different cases. The number of crews ranges from 50 in Case 1, with an increment of 25 in other cases, up to 150 for Case 5.

The model implemented on the IEEE 118-bus system setup comprises 162,481 decision variables, of which 22,308 are integer variables. The model also is constrained with 352,514 linear equations. The model is decomposed into an IP master problem and a dual linear subproblem and is solved using the Benders decomposition method. The optimal restoration schedule for buses and transmission lines

Table 10.4 Optimal repair schedule for buses in Scenarios I–III.

Bus #	Scenario I	Scenario II	Scenario III
B1	1–24	1–24	1–24
B2	1–11	1–11	1–11
B3	1–18	1–18	1–18
B4	1–15	1–15	1–15
B5	1–5	3–7	1–5
B8	1–4	1–4	6–9
B11	1–22	1–22	1–22

Table 10.5 Optimal repair schedule for transmission lines in Scenarios I–III.

Bus #	Scenario I	Scenario II	Scenario III
L1	5–24	84–103	92–111
L2	7–24	1–18	1–18
L10	6–21	19–34	95–100
L14	5–14	74–83	71–80
L16	1–22	28–49	48–69

of Scenarios I to III are shown in Tables 10.4 and 10.5, respectively. The costs of implementing Scenarios I and II, as well as the optimal restoration cost of the system with full consideration of the economics of disaster (Scenario III), are shown in Table 10.6. As shown, the implementation of Scenario I, which only minimizes the load interruption regardless of economic issues in the grid restoration process, results in a 12.3% increase in the cost of lost load as an index to measure social welfare. The overall restoration cost in this scenario increases by about 5% compared to Scenario III. In Scenario II, even though the objective function minimizes both lost load cost and repair crew cost, the induced impairment in the objective function results in an even higher cost of lost load and total restoration cost. Scenario III is a comprehensive economic restoration model, which is used as a benchmark for Scenarios I and II, and it provides the most economic restoration scheme and establishes an equilibrium between generation cost, lost load cost, and repair cost.

In Scenario IV, the economics of resources is analyzed. The optimal schedules for Cases 1–5 of Scenario IV's buses and lines are shown in Tables 10.7 and 10.8,

Table 10.6 Economic indices for Scenarios I–III (costs × 10^3).

Economic index	Scenario I	Scenario II	Scenario III
Total cost	\$30,188	\$30,261	\$28,764
Lost load cost	\$17,893	\$18,000	\$15,929
Generation cost	\$12,099	\$12,0067	\$12,640
Resource cost	\$19,692	\$19,476	\$19,362
Lost load (MWh)	2174	2275	3902

Table 10.7 Optimal repair schedule for buses in Scenario IV.

Bus #	$R_t = 50$	$R_t = 75$	$R_t = 100$	$R_t = 125$	$R_t = 150$
B1	27–50	1–24	1–24	1–24	1–24
B2	16–26	16–26	1–18	1–11	1–11
B3	28–45	25–42	1–18	1–18	1–18
B4	1–15	1–15	1–15	1–15	1–15
B5	1–5	1–5	1–5	3–7	4–8
B8	49–52	27–30	27–30	6–9	6–9
B11	6–27	1–22	1–22	1–22	1–22

Table 10.8 Optimal repair schedule for transmission lines in Scenario IV.

Line #	$R_t = 50$	$R_t = 75$	$R_t = 100$	$R_t = 125$	$R_t = 150$
L1	17–36	7–26	95–114	92–111	93–112
L2	59–76	47–64	6–23	1–18	1–18
L10	53–68	96–111	89–104	95–100	92–107
L14	1–10	25–34	71–80	71–80	71–80
L16	45–66	23–44	46–67	48–69	35–56

respectively. As expected, the load on B4, which is a critical load, is restored as quickly as possible in all scenarios. Due to their VOLLs, the repair operations on commercial and industrial loads are initiated from an early stage of the planning horizon to restore these costly interruptions. On the other hand, the restoration of the transmission lines in most cases is postponed to the middle or late stage of the planning horizon.

Table 10.9 Economic indices for different cases in Scenario IV (costs × 10^3).

Economic index	$R_t = 50$	$R_t = 75$	$R_t = 100$	$R_t = 125$	$R_t = 150$
Total cost	$37,713	$29,377	$28,798	$28,764	$28,762
Lost load cost	$24,952	$16,562	$15,966	$15,929	$15,926
Generation cost	$12,547	$12,598	$12,629	$12,640	$12,641
Resource cost	$21,324	$21,600	$20,346	$19,362	$19,410
Lost load (MWh)	6713	5713	4713	3713	3713
Interruption time (h)	50	42	24	24	24

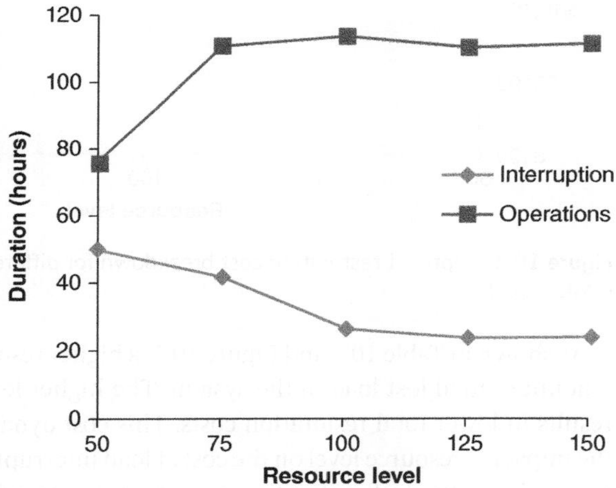

Figure 10.1 Interruption vs. operations duration for different cases in Scenario IV.

Table 10.9 summarizes the total restoration cost, the opportunity cost of the lost load, generation cost, resource cost, the amount of lost load, and the last time span that the system still experiences partial load interruption in Scenario IV. Due to the physics and economics of the problem, the restoration operations will continue for a longer period than the required time to eliminate partial load interruption in the system for each case, as shown in Figure 10.1. The higher level of restoration resources results in shorter interruption time in the system. Thus, the load interruption and operations duration diagram diverge by increasing the resource level, as illustrated in Figure 10.1. On the other hand, the higher resource level will extend the restoration planning horizon to complete the remaining operations in a cost-effective manner, i.e., by allocating less expensive resources and configuring a more economic generation unit commitment.

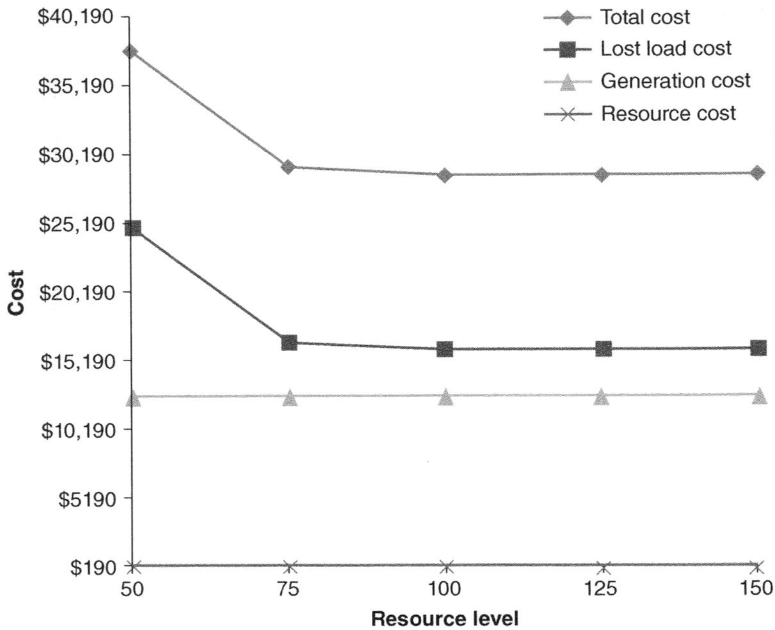

Figure 10.2 Optimal restoration cost breakdown for different cases in Scenario IV (costs × 10^3).

As shown in Table 10.9 and Figure 10.2, a higher resource level results in a lower amount of total lost load in the system. The higher level of restoration resources results in lower total restoration costs. This cost dynamics is significantly due to the impact of resource level on the cost of load interruption. Interestingly, by securing a higher level of restoration resources, the trend of the total resource cost shows a descending pattern. However, the total restoration resource cost is significantly lower than other cost components. At first glance, the generation cost is perceived to have low sensitivity to the resource level. However, considering different time-to-restoration lengths for different cases in Scenario IV, the average power generation per kWh for Cases 1–5 are $0.5663, $0.3848, $0.3721, $0.3847, and $0.3762, respectively. Therefore, the number of available restoration resources can significantly affect the average generation cost.

10.5 Summary

An efficient model was proposed to support the post-disaster decision-making process for power grid restoration. The results demonstrate that the proposed model can find the optimal restoration schedule for damaged components in a cost-effective manner. The opportunity cost of lost loads, the repair cost, and the generation costs were considered economic indices. It was demonstrated that the economy of disaster needs to be an important part of the restoration plan. Moreover, the numerical results showed that the restoration resource level significantly impacts the total incurred restoration cost. The results suggested that investing in restoration resources can be paid off by securing adequate restoration resources, resulting in considerable restoration cost savings. In addition, the higher level of resources can significantly shorten the partial grid restoration. It was shown that the number of available resources significantly impacts the average cost of power generation. This study suggests that incorporating the unit commitment problem, the VOLL, and repair crew cost into the restoration decision-making model can result in significant cost savings in the economic restoration of the grid.

References

1 J. Winkler, L. Dueñas-Osorio, R. Stein, and D. Subramanian, "Performance assessment of topologically diverse power systems subjected to hurricane events," *Rel. Eng. Syst. Safety*, vol. 95, no. 4, pp. 323–336, Apr. 2010.

2 H. E. Moore, F. L. Bates, M. V. Layman, and V. J. Parenton, "Before the wind: A study of the response to Hurricane Carla," Nat. Acad. Sci., Washington, DC, USA, Tech. Rep., 1963.

3 S. Guha, A. Moss, J. Naor, and B. Schieber, "Efficient recovery from power outage," in *Proc. 31th ACM Symp. Theory Comput.*, Atlanta, GA, USA, May 1999, pp. 574–582.

4 S. K. Khator and L. C. Leung, "Power distribution planning: A review of models and issues," *IEEE Trans. Power Syst.*, vol. 12, no. 3, pp. 1151–1159, Aug. 1997.

5 B. Ball, "Rebuilding electrical infrastructure along the Gulf coast: A case study," *Bridge-Washington-Nat. Acad. Eng.*, vol. 36, no. 1, p. 21, 2006.

6 A. Kwasinski, "Technology planning for electric power supply in critical events considering a bulk grid, backup power plants, and micro-grids," *IEEE Syst. J.*, vol. 4, no. 2, pp. 167–178, Jun. 2010.

7 N. Xu, S. D. Guikema, R. A. Davidson, L. K. Nozick, Z. Çanan, and K. Vaziri, "Optimizing scheduling of post-earthquake electric power restoration tasks," *Earthquake Eng. Struct. Dyn.*, vol. 36, no. 2, pp. 265–284, Feb. 2007.

8 D. Bienstock and S. Mattia, "Using mixed-integer programming to solve power grid blackout problems," *Discret. Optim.*, vol. 4, no. 1, pp. 115–141, Mar. 2007.

9 N. Perrier *et al.*, "A survey of models and algorithms for emergency response logistics in electric distribution systems. Part I: Reliability planning with fault considerations," *Comput. Oper. Res.*, vol. 40, no. 7, pp. 1895–1906, Jul. 2013.

10 N. Perrier *et al.*, "A survey of models and algorithms for emergency response logistics in electric distribution systems. Part II: Contingency planning level," *Comput. Oper. Res.*, vol. 40, no. 7, pp. 1907–1922, Jul. 2013.

11 M.-J. Yao and K. J. Min, "Repair-unit location models for power failures," *IEEE Trans. Eng. Manag.*, vol. 45, no. 1, pp. 57–65, Feb. 1998.

12 C. Coffrin, P. Van Hentenryck, and R. Bent, "Strategic stockpiling of power system supplies for disaster recovery," in *Proc. IEEE Power Eng. Soc. General Meeting*, San Diego, CA, USA, Jul. 2011, pp. 1–8.

13 Y. Yongbo *et al.*, "Service restoration with consideration of rush repair," in *Proc. Power Eng. Autom. Conf. (PEAM)*, Wuhan, China, Sep. 2011, pp. 308–312.

14 S. Wang, B. R. Sarker, L. Mann, Jr, and E. Triantaphyllou, "Resource planning and a depot location model for electric power restoration," *Eur. J. Oper. Res.*, vol. 155, no. 1, pp. 22–43, May 2004.

15 D. Lubkeman and D. E. Julian, "Large scale storm outage management," in *Proc. IEEE Power Eng. Soc. General Meeting*, Denver, CO, USA, Jun. 2004, pp. 16–22.

16 A. C. Chien, "Optimized recovery of damaged electrical power grids," M.S. thesis, Dept. Oper. Res., Naval Postgraduate School, Monterey, CA, USA, 2006.

17 S. Thiébaux, C. Coffrin, H. Hijazi, and J. Slaney, "Planning with MIP for supply restoration in power distribution systems," in *Proc. 23rd Int. Joint Conf. Artif. Intell.*, Beijing, China, Aug. 2013, pp. 2900–2907.

18 P. Van Hentenryck, N. Gillani, and C. Coffrin, "Joint assessment and restoration of power systems," in *Proc. 20th Euro. Conf. Artif. Intell. Appl.*, Montpellier, France, Aug. 2012, pp. 792–797.

19 T. C. Matisziw, A. T. Murray, and T. H. Grubesic, "Strategic network restoration," *Netw. Spat. Econ.*, vol. 10, no. 3, pp. 345–361, Sep. 2010.

20 S. G. Nurre, B. Cavdaroglu, J. E. Mitchell, T. C. Sharkey, and W. A. Wallace, "Restoring infrastructure systems: An integrated network design and scheduling (INDS) problem," *Eur. J. Oper. Res.*, vol. 223, no. 3, pp. 794–806, Dec. 2012.

21 P. Van Hentenryck, C. Coffrin, and R. Bent, "Vehicle routing for the last mile of power system restoration," in *Proc. 17th Power Syst. Comput. Conf.*, Stockholm, Sweden, Aug. 2011, pp. 1–9.

22 B. Simon, C. Coffrin, and P. Van Hentenryck, "Randomized adaptive vehicle decomposition for large-scale power restoration," in *Proc. 9th Int. Conf. Integr. AI OR Techn. Constraint Program. Combinat. Optim. Problems*, Nantes, France, Jun. 2012, pp. 379–394.

23 C. Coffrin, P. Van Hentenryck, and R. Bent, "Last-mile restoration for multiple interdependent infrastructures," in *Proc. 26th Conf. Artif. Intell.*, Toronto, ON, Canada, Jul. 2012, pp. 455–463.

24 X. Jiang, J. Chen, M. Chen and Z. Wei, "Multi-stage dynamic post-disaster recovery strategy for distribution networks considering integrated energy and transportation networks," *CSEE J. Power Energy Syst.*, vol. 7, no. 2, pp. 408–420, March 2021, doi: https://doi.org/10.17775/CSEEJPES.2020.02020.

25 H. Wu, Y. Xie, Y. Xu, Q. Wu, C. Yu, and J. Sun, "Robust coordination of repair and dispatch resources for post-disaster service restoration of the distribution system," *Int. J. Electr. Power Energy Syst.*, vol. 136, 2022, 107611.

26 Z. Ma, L. Zhang, Y. Cai, W. Tang, and C. Long, "Allocation method of coupled PV-energy storage-charging station in hybrid AC/DC distribution networks balanced with economics and resilience," *IET Renew. Power Gener.*, vol. 18, pp. 1060–1071, 2024. doi: https://doi.org/10.1049/rpg2.12864.

27 A. Arab, A. Khodaei, Z. Han, and S. K. Khator, "Proactive recovery of electric power assets for resiliency enhancement," *IEEE Access*, vol. 3, pp. 99–109, Mar. 2015.

28 A. Arab, A. Khodaei, S. K. Khator, K. Ding, V. A. Emesih, and Z. Han, "Stochastic pre-hurricane restoration planning for electric power systems infrastructure," *IEEE Trans. Smart Grid*, vol. 6, no. 2, pp. 1046–1054, Mar. 2015.

29 A. Arab, E. Tekin, A. Khodaei, S. K. Khator, and Z. Han, "Dynamic maintenance scheduling for power systems incorporating hurricane effects," in *Proc. IEEE Int. Conf. Smart Grid Commun.*, Venice, Italy, Nov. 2014, pp. 85–90.

30 CenterPoint Energy, *CenterPoint Energy*. 2013. Accessed: Jan 2023. [Online]. Available: http://www.centerpointenergy.com

31 J. Frayer, S. Keane, and J. Ng, "*Estimating the Value of Lost Load*," Boston, MA, USA: London Econ. Int. LLC, 2013.

32 M. Carrión and J. M. Arroyo, "A computationally efficient mixed-integer linear formulation for the thermal unit commitment problem," *IEEE Trans. Power Syst.*, vol. 21, no. 3, pp. 1371–1378, Aug. 2006.

33 M. P. Nowak and W. Römisch, "Stochastic Lagrangian relaxation applied to power scheduling in a hydro-thermal system under uncertainty," *Ann. Oper. Res.*, vol. 100, nos. 1–4, pp. 251–272, Dec. 2000.

34 A. Frangioni, C. Gentile, and F. Lacalandra, "Solving unit commitment problems with general ramp constraints," *Int. J. Electr. Power Energy Syst.*, vol. 30, no. 5, pp. 316–326, Jun. 2008.

35 A. J. Wood and B. F. Wollenberg, *Power Generation, Operation, and Control.* New York, NY, USA: Wiley, 2012.

36 M. Shahidehpour and Y. Fu, "Benders decomposition in restructured power systems," *IEEE Techtorial*, Apr. 2005.

37 E. Kalvelagen, "*Benders Decomposition with GAMS,*" Washington D C, USA: Amsterdam Optimization Modeling Group LLC, 2013. Accessed: Jan 2023. [Online]. Available: http://www.amsterdamoptimization.com/pdf/benders.pdf
38 U.S. Energy Information Administration, 2013. Accessed: Sep 2015. [Online]. Available: http://www.eia.gov.

Index

Note: *Italicized* and **bold** page numbers refer to figures and tables, respectively.

d

q

r

Printed and bound by CPI Group (UK) Ltd, Croydon, CR0 4YY
28/07/2026
14931303-0001